Weighted Inequalities Involving ρ-quasiconcave Operators

Weighted Inequalities Involving ρ-quasiconcave Operators

William Desmond Evans
School of Mathematics
Cardiff University
Wales, U.K.

Amiran Gogatishvili
Institute of Mathematics
Czech Academy of Sciences
Czech Republic

Bohumír Opic
Faculty of Mathematics and Physics
Charles University
Czech Republic

NEW JERSEY • LONDON • SINGAPORE • BEIJING • SHANGHAI • HONG KONG • TAIPEI • CHENNAI • TOKYO

Published by

World Scientific Publishing Co. Pte. Ltd.
5 Toh Tuck Link, Singapore 596224
USA office: 27 Warren Street, Suite 401-402, Hackensack, NJ 07601
UK office: 57 Shelton Street, Covent Garden, London WC2H 9HE

Library of Congress Cataloging-in-Publication Data
Names: Evans, W. D., author. | Gogatishvili, Amiran, author. | Opic, B. (Bohumír), author.
Title: Weighted inequalities involving ρ-quasiconcave operators / by William Desmond Evans (Cardiff University, UK), Amiran Gogatishvili (Academy of Sciences of the Czech Republic, Czech Republic), Bohumír Opic (Charles University, Czech Republic).
Description: New Jersey : World Scientific, 2018. | Includes bibliographical references and index.
Identifiers: LCCN 2018011306 | ISBN 9789813239623 (hardcover : alk. paper)
Subjects: LCSH: Nonlinear operators. | Operator theory. | Functional analysis. | Vector spaces.
Classification: LCC QA329.8 .E93 2018 | DDC 515/.88--dc23
LC record available at https://lccn.loc.gov/2018011306

British Library Cataloguing-in-Publication Data
A catalogue record for this book is available from the British Library.

For any available supplementary material, please visit
https://www.worldscientific.com/worldscibooks/10.1142/10974#t=suppl

Printed in Singapore

To Mari, Ia and Monika

Contents

Acknowledgement

The research was supported by grants nos. 201/01/0333, 201/05/2033, 201/08/0383 and 201/13/14743S of the Grant Agency of the Czech Republic, by the Leverhulme Trust grant no. F/00407/E, and by the INTAS grant no. 8157.

The research of A. Gogatishvili was also supported by the RVO grant no. 67985840 and partially supported by the SRNSF grant no. 217282.

Preface

The main theme of these notes is a detailed study of the class $Q_\rho(I)$ of *ρ-quasiconcave functions* defined on an interval $I = (a, b) \subseteq \mathbb{R}$, and the application of the results established to characterize the validity of weighted Hardy-type inequalities. The function ρ is positive, continuous and strictly increasing on I, and a function h belongs to $Q_\rho(I)$ if it is non-negative, and both h and ρ/h are non-decreasing on I. An important subclass $Q^0_\rho(I)$ of $Q_\rho(I)$ consists of functions which are such that

$$h(t), \big(\rho/h\big)(t) \to 0 \quad as \quad t \to a+ \quad \text{and} \quad h(t), \big(\rho/h\big)(t) \to +\infty \quad as \quad t \to b-\,.$$

When ρ is the identity, we simply write $Q(I)$ and $Q^0(I)$ instead of $Q_\rho(I)$ and $Q^0_\rho(I)$, respectively. If $I = (0, +\infty)$, $Q(I)$ is the class of quasiconcave functions. The classes $Q((0, +\infty))$, $Q^0((0, +\infty))$ have been extensively studied and applied to a variety of problems; see, for example, [Osk1], [Osk2], [BK, chapter 3], [J], [C], [Ov1], [Ov2], [BrSh], [N], [BeSha] and [GHS]. They have a role of particular importance in interpolation theory. In [J] Janson developed new interpolation methods based on quasiconcave functions, and such functions were also the basis of Ovchinnikov's generalization in [Ov2] of Janson's method with respect to a class Φ of interpolation functions $\varphi : [0, \infty) \times [0, \infty) \to [0, \infty)$ with the properties $\varphi(s, t) > 0$, $\varphi(\cdot, t)$ and $\varphi(s, \cdot)$ increasing, and $\varphi(\lambda s, \lambda t) = \lambda\varphi(s, t)$ for all $\lambda, s, t > 0$; each $\varphi \in \Phi$ can be shown to be quasiconcave in each variable separately. Quasiconcave functions also feature prominently in the proofs of the fundamental lemma of interpolation and the Brudnyĭ-Krugljak K-divisibility theorem; see [BeSha, Chapter 5].

For instances of work on $Q^0_\rho(I)$ in which ρ is not the identity function, see [G1], [G2] and [P], where $\rho(t) = t^k$, $k > 0$, and $I = (0, +\infty)$, $I = (0, +\infty)$ and $I = (1, +\infty)$, respectively; [G1] and [G2] are concerned with integral inequalities on a cone of quasiconcave functions, while in [P], the results obtained are applied to investigate relationships between the summability of functions and their Fourier series. In [GP1], [GP2] and [GJOP], the class $Q^0((0, +\infty))$ is used to obtain integral conditions characterizing embeddings between Lorentz spaces $\Lambda^q(w)$ and $\Gamma^p(v)$, and to determine an integral description of the associate space to $\Gamma^p(v)$.

Some problems in mathematical analysis (e.g., in the theory of function spaces, approximation theory and interpolation theory) lead to the investigation of weighted inequalities containing certain classes of quasiconcave functions on an interval $I = (a,b) \subseteq \mathbb{R}$. Thus, in this book we analyze the class $Q_\rho(\mathrm{I})$ of ρ-quasiconcave functions in complete generality, in order to establish results needed for a comprehensive study of the general Hardy-type inequalities and their reverse form made in the final chapter. An important result is that to any $h \in Q_\rho(I)$, a sequence $\{x_k\}$ of points in the interval I can be assigned which may be used to decompose I into a system $\{I_k\}$ of disjoint sub-intervals I_k with respect to which

$$\text{(0.0.1)} \qquad \text{either } h(x) \approx h(y), \text{ or } \frac{h}{\rho}(x) \approx \frac{h}{\rho}(y) \text{ for all } x, y \in I_k,$$

where the notation $\approx$ indicates that the quotient of the two sides is bounded below and above on I_k by positive absolute constants; the sequence is said to be a *ρ-discretizing sequence* of h. The properties of $Q_\rho(I)$ provide us with the means to make a thorough investigation of inequalities like

$$\text{(0.0.2)} \qquad \left\| \sup_{a<t\leq x} \rho(t) \int_t^b g(s)\,ds \right\|_{q,w,I} \lesssim \|g\|_{p,v,I},$$

for all non-negative Lebesgue measurable functions g on the interval I, where w, v are weights on I (we refer to Section 1.1 below for the definition of weighted (quasi)-norms $\|\cdot\|_{q,w,I}$ and $\|\cdot\|_{p,v,I}$). Necessary and sufficient conditions for (0.0.2) to hold are given for the full range of indices, $0 < p \leq +\infty$, $0 < q \leq +\infty$, with the understanding that when $0 < p \leq 1$, the inequality in (0.0.2) is reversed. The results are obtained by the technique of *discretization/anti-discretization.* The discretization stage involves expressing the weighted quasi-norms on both sides of the inequality as series, using the ρ-discretizing sequences of ρ-quasiconcave functions present, and then resolving inequality (0.0.2) at the level of each integer (involved in the problem) by means of known criteria. The local problem is an easier one as it involves one operation less. This technique was developed in [GHS], and has proved to be an effective tool in many problems. However, a drawback is that the discrete criteria determined for the inequality under consideration are often very difficult to verify, and this is what motivated the development of the anti-discretization technique, which converts the discrete criteria into continuous form. The discretization/anti-discretization method developed here is modeled on that in [GP1] with respect $Q_\rho^0((0,+\infty))$. The desire to have available results for general $Q_\rho(I)$ functions, so that attempting to adapt ones for $Q_\rho^0((0,+\infty))$ would be unnecessary, was a reason for undertaking the present study. What makes it possible for us to apply this technique is the body of results on the properties of $Q_\rho(I)$ proved earlier, that yield equivalent discrete and continuous expressions for the quasi-norms of various quasi-Banach spaces.

Another inequality treated in full is

$$\text{(0.0.3)} \qquad \left\| \int_a^x u(t) \left(\int_t^b g(s)\,ds \right) dt \right\|_{q,w,I} \lesssim \|g\|_{p,v,I}$$

for all non-negative Lebesgue measurable functions g on the interval I, where u, v, w are weights on I, and the full range $0 < p, q \leq +\infty$, is again considered, with the inequality reversed when $0 < p \leq 1$. Inequalities (0.0.2) and (0.0.3) are typical examples of inequalities that are amenable to our approach and that play an important role in mathematical analysis.

After introducing the notation and discussing briefly some preliminary material on Lebesgue-Stieltjes and Riemann-Stieltjes integrals in Chapter 1, we embark on a detailed study of the class $Q_\rho(I)$ in Chapter 2. With applications in mind, we construct a lengthy list of ρ-quasiconcave functions with respect to a given *admissible* function ρ, that is, a function which is positive, continuous and strictly increasing on I. A particularly important example for our needs is the following:

$$\varphi(x) := \| \min\{\rho(\cdot), \rho(x)\} \|_{p,w,I,\mu} \tag{0.0.4}$$

where $p \in (0, +\infty]$, w is a weight on the interval I, and μ a non-negative Borel measure on I. This function φ is called the *ρ-fundamental function* of the weighted Lebesgue space $L^p(w, I, \mu)$ with (quasi)-norm $\|\cdot\|_{p,w,I,\mu}$ given in (1.1.1) below. It is also proved in Chapter 2 that any $h \in Q_\rho(I)$ can be represented by means of a non-negative Borel measure μ on I in such a way that, for all $x \in I$,

$$h(x) \leq \alpha + \beta\rho(x) + \int_I \min\{\rho(x), \rho(t)\} d\mu(t) \leq 4h(x), \tag{0.0.5}$$

and hence

$$h(x) \approx \alpha + \beta\rho(x) + \int_I \min\{\rho(x), \rho(t)\} d\mu(t),$$

where

$$\alpha = \lim_{t \to a+} h(t), \quad \beta = \lim_{t \to b-} \frac{h(t)}{\rho(t)}.$$

This generalizes a well-known result on the representation of functions in $Q((0, +\infty))$, see for example [BL, page 117]. Also, for any $p \in (0, +\infty)$, h^p and $(\rho/h)^p$ belong to $Q_{\rho^p}(I)$, and the representations of type (0.0.5) corresponding to these functions are determined in terms of the representation for h. Furthermore, it is proved that if h is the ρ-fundamental function of $L^1(w, I, \mu)$, then there is a weight W_p, $p \in (0, +\infty]$, such that h is equivalent to the ρ-fundamental function of $L^p(W_p, I, \mu)$ and also that

$$\left(\frac{\rho}{h}\right)(x) \approx \| \min\{\rho(\cdot), \rho(x)\} \|_{\infty,1/h,I,\mu}. \tag{0.0.6}$$

The existence and some properties of a discretizing sequence, for any $h \in Q_\rho(I)$, are proved in Chapter 3, and this information is used in Chapter 4 to discretize weighted quasi-norms of various ρ-quasiconcave functions. The material in Chapter 4 provides the tools for use in applying the discretization/anti-discretization method to the integral inequalities in the next chapter.

The effectiveness of the method in establishing necessary and sufficient conditions for the validity of inequalities like (0.0.2) and (0.0.3) is demonstrated clearly in the final chapter. The resolution of the discrete inequality that appears in the first step is relatively straightforward. The difficulties occur in the second step, in which the discrete criteria obtained in the first step are to be translated into a continuous form, using results established largely in Chapter 4. Other techniques can be modified to treat the inequalities like (0.0.2) and (0.0.3) if $1 \le p \le +\infty$ (in the case of inequality (0.0.2) see, e.g., [GOP], [K], [Pr1], [Pr2], [PS1], and [PS2]; in the case of inequality (0.0.3) methods of [OK] or [KP] can be applied).[1] But these can not be used for the reverse inequalities, when $0 < p \le 1$. For the latter, the method presented in this book is particularly successful in combination with results on reverse inequalities obtained in [EGO1]. This method was used in [GMP1] and [GMP2] to investigate more general inequalities than (0.0.3).

Some results in this book were presented at the Conference on Inequalities and Applications, Noszvaj (Hungary), September 2007; see [EGO2].

[1]Concerning inequality (0.0.3) note that the Fubini theorem implies

$$\Big\| \int_a^x u(t) \Big(\int_t^b g(s)\,ds \Big)\,dt \Big\|_{q,w,I} \approx \Big\| \int_a^x g(s) U(s)\,ds \Big\|_{q,w,I} + \Big\| U(x) \int_x^b g(s)\,ds \Big\|_{q,w,I},$$

where $U(x) := \int_a^x u(t)\,dt$, $x \in I$, for all functions g in question.

Basic Notation

$\mathbb{R}$: real numbers.

$\mathbb{R}^n$: n-dimensional Euclidean spaces.

$\mathbb{C}$: complex numbers.

$\mathbb{N}$: natural numbers.

$\mathbb{N}_0 = \mathbb{N} \cup \{0\}$.

$\mathbb{Z}$: integer numbers.

$\overline{\mathbb{Z}} = \mathbb{Z} \cup \{-\infty, +\infty\}$.

$X \hookrightarrow Y$: $X \subset Y$ and the natural embedding of X in Y is continuous.

$A \lesssim B$: (or $A \gtrsim B$) if $A \leq cB$ (or $cA \geq B$) for some positive constant c independent of appropriate quantities involved in the expressions A and B.

$A \approx B$: $A \lesssim B$ and $A \gtrsim B$.

LHS(*) (RHS(*)) for the left- (right-) hand side of the relation (*).

χ_Q: the characteristic function of a set Q.

$p' = \frac{p}{p-1}$ when $p \in (1, +\infty)$, $p' = +\infty$ if $p = 1$ and $p' = 1$ if $p = +\infty$.

$p^* = \frac{p}{1-p}$ when $p \in (0, 1)$ and $p^* = +\infty$ if $p = 1$.

$r_+ = \max\{r, 0\}$ when $r \in \mathbb{R}$.

Chapter 1

Preliminaries

1.1 Notation

Throughout these notes we assume that $I := (a,b) \subseteq \mathbb{R}$. Let μ be a non-negative measure on I. By $\mathcal{M}(I,\mu)$ we denote the set of all μ-measurable functions on I. The symbol $\mathcal{M}^+(I,\mu)$ stands for the collection of all $f \in \mathcal{M}(I,\mu)$ which are non-negative on I, while $\mathcal{M}^+(I,\mu;\uparrow)$ or $\mathcal{M}_s^+(I,\mu;\uparrow)$ is used to denote the subset of those functions which are non-decreasing or (strictly) increasing, respectively, on I. The sets $\mathcal{M}^+(I,\mu;\downarrow)$ and $\mathcal{M}_s^+(I,\mu;\downarrow)$ are defined analogously. The family $\mathcal{W}(I,\mu)$ of all *weight functions* on I is given by

$$\mathcal{W}(I,\mu) = \{w \in \mathcal{M}(I,\mu);\ w > 0\ \mu\text{-a.e. on } I\}.$$

If the measure μ is the Lebesgue measure on I, then we omit the symbol μ in the notation and, for example, we write simply $\mathcal{M}(I)$ instead of $\mathcal{M}(I,\mu)$.

The symbol $Ads(I)$ stands for the set of all *admissible functions* on I given by

$$Ads(I) = \mathcal{M}_s^+(I;\uparrow) \cap C(I),$$

where $C(I)$ denotes the family of all continuous functions on I.

For $p \in (0,+\infty]$ and $w \in \mathcal{M}^+(I,\mu)$, we define the functional $\|\cdot\|_{p,w,I,\mu}$ on $\mathcal{M}(I,\mu)$ by [1]

$$\|f\|_{p,w,I,\mu} = \begin{cases} \left(\int_I |fw|^p\, d\mu\right)^{1/p} & \text{if } p < +\infty \\ \operatorname*{ess\,sup}_I |fw| & \text{if } p = +\infty. \end{cases} \tag{1.1.1}$$

If, in addition, $w \in \mathcal{W}(I,\mu)$, then the *weighted Lebesgue space* $L^p(w,I,\mu)$ is given by

$$L^p(w,I,\mu) = \{f \in \mathcal{M}(I,\mu);\ \|f\|_{p,w,I,\mu} < +\infty\} \tag{1.1.2}$$

[1] When E is a μ–measurable subset of I, then the functional $\|\cdot\|_{p,w,E,\mu}$ is defined analogously to (1.1.1).

and it is equipped with the quasi-norm $\|\cdot\|_{p,w,I,\mu}$.

When $w \equiv 1$ on I, we write simply $L^p(I,\mu)$ and $\|\cdot\|_{p,I,\mu}$ instead of $L^p(w,I,\mu)$ and $\|\cdot\|_{p,w,I,\mu}$, respectively. Furthermore, if μ is the Lebesgue measure, then we use symbols $L^p(I)$, $\|\cdot\|_{p,I}$, $L^p(w,I)$ and $\|\cdot\|_{p,w,I}$ instead of $L^p(I,\mu)$, $\|\cdot\|_{p,I,\mu}$, $L^p(w,I,\mu)$ and $\|\cdot\|_{p,w,I,\mu}$, respectively.

Finally, if $p \in (0,+\infty]$, $\mathcal{Z} \subseteq \mathbb{Z}$, $\mathcal{Z} \neq \emptyset$ and $\{w_k\} = \{w_k\}_{k\in\mathcal{Z}}$ is a sequence of positive numbers, then the discrete analogue of $L^p(w,I)$ is denoted by $\ell^p(\{w_k\},\mathcal{Z})$ and, when $w_k = 1$ for all $k \in \mathcal{Z}$, simply by $\ell^p = \ell^p(\mathcal{Z})$. Sometimes we write $\|a_k\|_{\ell^p(\mathcal{Z})}$ instead of $\|\{a_k\}\|_{\ell^p(\mathcal{Z})}$.

Given two quasi-Banach spaces X and Y, we write $X = Y$ if X and Y are equal in the algebraic and the topological sense (their quasi-norms are equivalent). The symbol $X \hookrightarrow Y$ means that $X \subset Y$ and the natural embedding of X in Y is continuous.

We write $A \lesssim B$ (or $A \gtrsim B$) if $A \le cB$ (or $cA \ge B$) for some positive constant c independent of appropriate quantities involved in the expressions A and B, and $A \approx B$ (and say that A *is equivalent to* B) if $A \lesssim B$ and $A \gtrsim B$. Throughout these notes we use the abbreviation LHS($*$) (RHS($*$)) for the left- (right-) hand side of the relation ($*$). By χ_Q we denote the characteristic function of a set Q.

Convention 1.1.1 (i) Throughout these notes we put $1/(+\infty) = 0$, $(+\infty)/(+\infty) = 0$, $1/0 = (+\infty)$, $0/0 = 0$, $0\cdot(\pm\infty) = 0$, $(+\infty)^\alpha = +\infty$ and $\alpha^0 = 1$ if $\alpha \in (0,+\infty)$.
(ii) If $p \in [1,+\infty]$, we define p' by $1/p + 1/p' = 1$. Moreover, we put $p^* = \frac{p}{1-p}$ if $p \in (0,1]$.
(iii) If $r \in \mathbb{R}$, then we define r_+ by $r_+ = \max\{r,0\}$.
(iv) If $I = (a,b) \subseteq \mathbb{R}$ and $g \in \mathcal{M}^+(I;\uparrow)$ or $g \in \mathcal{M}^+(I;\downarrow)$, then by $g(a)$ and $g(b)$ we mean the limits $\lim_{x\to a_+} g(x)$ and $\lim_{x\to b_-} g(x)$, respectively.

1.2 Some results on measures and integration

The symbol $\mathcal{B}^+(I)$ stands for the collection of all non-negative Borel measures on the interval I. If $\mu \in \mathcal{B}^+(I)$, then *the extension of μ by zero in $\mathbb{R}\setminus I$* is the measure $\bar{\mu} \in \mathcal{B}^+(\mathbb{R})$ defined by $\bar{\mu}(S) := \mu(S\cap I)$ for any Borel set $S \subset \mathbb{R}$. If $x_1, x_2 \in I$, $x_1 < x_2$, then by $AC([x_1,x_2])$ we mean the set of all real functions which are absolutely continuous on $[x_1,x_2]$.

Let φ be *non-decreasing* and finite function on the interval $I := (a,b) \subseteq \mathbb{R}$. We assign to φ the function λ defined on subintervals of I by

$$\lambda([y,z]) = \varphi(z+) - \varphi(y-), \tag{1.2.1}$$

$$\lambda([y,z)) = \varphi(z-) - \varphi(y-), \tag{1.2.2}$$

$$\lambda((y,z]) = \varphi(z+) - \varphi(y+), \tag{1.2.3}$$

$$\lambda((y,z)) = \varphi(z-) - \varphi(y+). \tag{1.2.4}$$

The function λ is a non-negative, additive and regular function of intervals. Thus (cf. [R, Chapter 10]), it admits a unique extension to a non-negative Borel measure λ on I.

Note also that the associated Borel measure can be determined, e.g., only by putting

$$\lambda([y,z]) = \varphi(z+) - \varphi(y-) \quad \text{for any} \quad [y,z] \subset I$$

(since the Borel subsets of I can be generated by subintervals $[y,z] \subset I$).

If $J \subseteq I$, then the Lebesgue-Stieltjes integral $\int_J f\,d\varphi$ is defined as $\int_J f\,d\lambda$. We shall also use the Lebesgue-Stieltjes integral $\int_J f\,d\varphi$ when φ is a *non-increasing* and finite on the interval I. In such a case we put

$$\int_J f\,d\varphi := -\int_J f\,d(-\varphi).$$

Remark 1.2.1 Let $I = (a,b) \subseteq \mathbb{R}$. If $f \in C(I)$ and φ is a non-decreasing, right continuous and finite function on I, then it is possible to show that, for any $[y,z] \subset I$, the Riemann-Stieltjes integral $\int_{[y,z]} f\,d\varphi$ (written usually as $\int_y^z f\,d\varphi$) coincides with the Lebesgue-Stiltjes integral $\int_{(y,z]} f\,d\varphi$. In particular, if $f, \varphi \in C(I)$ and φ is non-decreasing on I, then the Riemann-Stieltjes integral $\int_{[y,z]} f\,d\varphi$ coincides with the Lebesgue-Stieltjes integral $\int_{[y,z]} f\,d\varphi$ for any $[y,z] \subset I$.

In Chapter 2 we shall use the following assertions involving the Lebesgue-Stieltjes integrals.

Lemma 1.2.2 *Let $I = (a,b) \subseteq \mathbb{R}$. Assume that φ, ψ are non-decreasing on I, $\varphi \in C(I)$ and that ψ is finite and right continuous on I. Define the functions ν and μ by*

$$\nu([y,z]) = \int_{[y,z]} \varphi\,d\psi + \int_{[y,z]} \psi\,d\varphi \tag{1.2.5}$$

and

$$\mu([y,z]) = \int_{[y,z]} d(\varphi\psi) \tag{1.2.6}$$

for any interval $[y,z] \subset I$. Then ν and μ can be uniquely extended to non-negative Borel measures on I such that, for any Borel subset $S \subseteq I$,

$$\nu(S) = \mu(S).$$

Proof. It is clear that ν and μ are additive functions of intervals. Thus, if we show that

$$\nu([y,z]) = \mu([y,z]) \tag{1.2.7}$$

for any $[y,z] \subset I$, then the result follows.

Let $[y, z] \subset I$. Then

$$\mu([y, z]) = (\varphi\psi)(z+) - (\varphi\psi)(y-) = (\varphi\psi)(z) - (\varphi\psi)(y-). \tag{1.2.8}$$

On the other hand, since $\varphi \in C(I)$ and ψ is right continuous on I,

$$\int_{[y,z]} \varphi \, d\psi = \int_{(y,z]} \varphi \, d\psi + \varphi(y)[\psi(y) - \psi(y-)] \tag{1.2.9}$$

and

$$\int_{[y,z]} \psi \, d\varphi = \int_{(y,z]} \psi \, d\varphi. \tag{1.2.10}$$

Moreover, integration by parts (cf., e.g., [Fo, Theorem 3.36]) yields

$$\int_{(y,z]} \varphi \, d\psi + \int_{(y,z]} \psi \, d\varphi = (\varphi\psi)(z) - (\varphi\psi)(y). \tag{1.2.11}$$

By (1.2.5) and (1.2.9)-(1.2.11),

$$\nu([y, z]) = (\varphi\psi)(z) - (\varphi\psi)(y) + \varphi(y)[\psi(y) - \psi(y-)] = (\varphi\psi)(z) - (\varphi\psi)(y-).$$

Together with (1.2.8), this implies (1.2.7). □

Lemma 1.2.3 *Let $I = (a, b) \subseteq \mathbb{R}$. Assume that φ, ψ are non-decreasing on I, $\varphi \in C(I)$ and that ψ is finite and left continuous on I. Define the functions ν and μ by (1.2.5) and (1.2.6) for any interval $[y, z] \subset I$. Then ν and μ can be uniquely extended to non-negative Borel measures on I such that, for any Borel subset $S \subseteq I$,*

$$\nu(S) = \mu(S).$$

Proof is analogous to that of Lemma 1.2.2.

Lemma 1.2.4 *Let $\alpha \neq 0$ and let φ be a positive, continuous and strictly monotone function on $I = (a, b) \subseteq (0, +\infty)$. Define the functions ν and μ by*

$$\nu([y, z]) = \int_{[y,z]} d[\varphi^{\alpha}(t)]$$

and

$$\mu([y, z]) = \int_{[y,z]} \alpha\varphi^{\alpha-1}(t) \, d\varphi(t)$$

for any interval $[y, z] \subset I$. Then ν and μ can be uniquely extended to signed Borel measures[2] on I such that, for any Borel subset $S \subseteq I$,

$$\nu(S) = \mu(S).$$

[2] If φ^{α} is increasing on I, then $\mu, \nu \in \mathcal{B}^{+}(I)$. When φ^{α} is decreasing on I, then $(-\mu), (-\nu) \in \mathcal{B}^{+}(I)$.

Proof. Let $[y,z] \subset I$. Then

$$\nu([y,z]) = \int_{[y,z]} d[\varphi^\alpha(t)] = \varphi^\alpha(z) - \varphi^\alpha(y). \tag{1.2.12}$$

By Remark 1.2.1, the Lebesgue-Stieltjes integral defining the function μ coincides with the corresponding Riemann-Stieltjes one since $\varphi \in C(I)$. Using this fact and a change of variables (cf., e.g., [R, Theorem 6.33] or [Fe, 2.5.18 (2)]), we obtain

$$\mu([y,z]) = \int_y^z \alpha\varphi^{\alpha-1}(t)\, d\varphi(t) = \int_{\varphi(y)}^{\varphi(z)} \alpha s^{\alpha-1}\, ds. \tag{1.2.13}$$

Since

$$\int_{\varphi(y)}^{\varphi(z)} \alpha s^{\alpha-1}\, ds = \varphi^\alpha(z) - \varphi^\alpha(y) = \nu([y,z]),$$

the result follows. □

Lemma 1.2.5 *Let $p \in (0,+\infty)$, $I = (a,b) \subseteq \mathbb{R}$, and let $\lambda \in \mathcal{B}^+(I)$ satisfy*

$$L(x) := \int_{(a,x]} d\lambda < +\infty \quad \textit{for all} \quad x \in I.$$

Then, for all $x \in I$,

$$\min\{1, \frac{1}{p}\}\, L^p(x) \le \int_{(a,x]} L^{p-1}\, d\lambda \le \max\{1, \frac{1}{p}\}\, L^p(x). \tag{1.2.14}$$

Proof is analogous to that of [OPS, Lemma 2], where $I = \mathbb{R}$. □

Lemma 1.2.6 *Let $p \in (0,+\infty)$, $I = (a,b) \subseteq \mathbb{R}$, and let $\mu \in \mathcal{B}^+(I)$ satisfy*

$$M(x) := \int_{[x,b)} d\mu < +\infty \quad \textit{for all} \quad x \in I.$$

Then, for all $x \in I$,

$$\min\{1, \frac{1}{p}\}\, M^p(x) \le \int_{[x,b)} M^{p-1}\, d\mu \le \max\{1, \frac{1}{p}\}\, M^p(x). \tag{1.2.15}$$

Proof is analogous to that of [OPS, Corollary 1], where $I = \mathbb{R}$. □

Lemma 1.2.7 *Let $I = (a,b) \subseteq \mathbb{R}$ and let $\lambda \in \mathcal{B}^+(I)$ satisfy*

$$L(x) := \int_{(a,x]} d\lambda < +\infty \quad \textit{for all} \quad x \in I.$$

Take $\alpha \in \mathbb{R}$ and define the function ν by

$$\nu([y,z]) = \int_{[y,z]} d(\alpha + L)$$

for any interval $[y,z] \subset I$. Then ν can be uniquely extended to a non-negative Borel measure on I which coincides with λ.

Proof. Since the function L is non-decreasing, finite and right continuous on I, the function $\varphi := \alpha + L$ has the same properties. Moreover,

$$L(x-) = \int_{(a,x)} d\lambda \quad \text{for all} \quad x \in I.$$

Consequently, for any interval $[y,z] \subset I$,

$$\begin{aligned}\nu([y,z]) &= \int_{[y,z]} d\varphi = \varphi(z+) - \varphi(y-) = L(z) - L(y-) \\ &= \int_{(a,z]} d\lambda - \int_{(a,y)} d\lambda = \int_{[y,z]} d\lambda = \lambda([y,z])\end{aligned}$$

and the result follows. □

Lemma 1.2.8 *Let $I = (a,b) \subseteq \mathbb{R}$ and $\mu \in \mathcal{B}^+(I)$ satisfy*

$$M(x) := \int_{[x,b)} d\mu < +\infty \quad \textit{for all} \quad x \in I.$$

Take $\beta \in \mathbb{R}$ and define the function ν by

$$\nu([y,z]) = -\int_{[y,z]} d(\beta + M)$$

for any interval $[y,z] \subset I$. Then ν can be uniquely extended to a non-negative Borel measure on I which coincides with μ.

Proof. Since the function M is non-increasing, finite and left continuous on I, the function $\varphi := -\beta - M$ is non-decreasing, finite and left continuous on I. Moreover,

$$M(x+) = \int_{(x,b)} d\mu \quad \text{for all} \quad x \in I.$$

Consequently, for any interval $[y,z] \subset I$,

$$\begin{aligned}\nu([y,z]) &= \int_{[y,z]} d\varphi = \varphi(z+) - \varphi(y-) = M(y) - M(z+) \\ &= \int_{[y,b)} d\mu - \int_{(z,b)} d\mu = \int_{[y,z]} d\mu = \mu([y,z])\end{aligned}$$

and the result follows. □

Corollary 1.2.9 *Let ν and μ be Borel measures from* Lemma 1.2.2 *(or* Lemma 1.2.3 *or* Lemma 1.2.4 *or* Lemma 1.2.8*). Then*

$$\int_S f\, d\nu = \int_S f\, d\mu$$

whenever one of the integrals exists.

Similarly, if λ and ν are Borel measures from Lemma 1.2.7, *then*

$$\int_S f\,d\lambda = \int_S f\,d\nu$$

whenever one of the integrals exists.

Remark 1.2.10 As usual, if ν and μ are measures from Lemma 1.2.2, then we write

$$\varphi\,d\psi + \psi\,d\varphi \quad \text{and} \quad d(\varphi\psi)$$

instead of $d\nu$ and $d\mu$, respectively. Similarly we proceed in other cases. For example, if ν and μ are measures from Lemma 1.2.4, then we write

$$d\varphi^\alpha \quad \text{and} \quad \alpha\varphi^{\alpha-1}\,d\varphi$$

instead of $d\nu$ and $d\mu$.

1.3 Almost geometric increasing or decreasing sequences

Definition 1.3.1 Let $L, M \in \overline{\mathbb{Z}}$, $L < M$. A positive non-increasing sequence $\{\tau_k\}_{k=L}^M$ is said to be *almost geometrically decreasing* if there is $\alpha \in (1, +\infty)$ and $n \in \mathbb{N}$ such that

$$\tau_k \leq \frac{1}{\alpha}\tau_{k-n} \quad \text{for all} \quad k \in \mathbb{Z},\ L+n \leq k \leq M.$$

In the case $n = 1$ the sequence is said to be *geometrically decreasing.*

A positive non-decreasing sequence $\{\sigma_k\}_{k=L}^M$ is said to be *almost geometrically increasing* if there is $\alpha \in (1, +\infty)$ and $n \in \mathbb{N}$ such that

$$\sigma_k \geq \alpha\sigma_{k-n} \quad \text{for all} \quad k \in \mathbb{Z},\ L+n \leq k \leq M.$$

In the case $n = 1$ the sequence is said to be *geometrically increasing.*

Remark 1.3.2 Definition 1.3.1 implies that if $0 < q < +\infty$, then following three statements are equivalent:

(i) $\{\tau_k\}_{k=N}^M$ is an almost geometrically decreasing sequence;
(ii) $\{\tau_k^q\}_{k=N}^M$ is an almost geometrically decreasing sequence;
(iii) $\{\tau_k^{-q}\}_{k=N}^M$ is an almost geometrically increasing sequence.

The equivalence continues to be true if the word *almost* is deleted from each statement.

In Chapters 4 and 5 we shall need the following assertions, whose proofs can be found in [L1], [L2].

Lemma 1.3.3 *Let $L, M \in \overline{\mathbb{Z}}$, $L < M$. Then, for any positive sequence $\{\tau_k\}_{k=L}^{M}$,*

$$\sum_{k=m}^{M} \tau_k \lesssim \tau_m \qquad \text{for all} \quad m \in \mathbb{Z},\ L < m \leq M, \tag{1.3.1}$$

or

$$\sum_{k=L}^{m} \tau_k \lesssim \tau_m \qquad \text{for all} \quad m \in \mathbb{Z},\ L \leq m < M, \tag{1.3.2}$$

if and only if the sequence $\{\tau_k\}_{k=L}^{M}$ is an almost geometrically decreasing or increasing, respectively.

Lemma 1.3.4 *Let $q \in (0, +\infty]$, $L, M \in \overline{\mathbb{Z}}$, $L < M$, $S = \{k \in \mathbb{Z};\ L \leq k \leq M\}$ and let $\{\tau_k\}_{k=L}^{M}$ be an almost geometrically decreasing sequence. Then*

$$\left\| \tau_k \sum_{m=L}^{k} a_m \right\|_{\ell^q(S)} \approx \|\tau_k a_k\|_{\ell^q(S)} \tag{1.3.3}$$

and

$$\|\tau_k \sup_{L \leq m \leq k} a_m\|_{\ell^q(S)} \approx \|\tau_k a_k\|_{\ell^q(S)} \tag{1.3.4}$$

for all non-negative sequences $\{a_k\}_{k=L}^{M}$.

Lemma 1.3.5 *Let $q \in (0, +\infty]$, $L, M \in \overline{\mathbb{Z}}$, $L < M$, $S = \{k \in \mathbb{Z};\ L \leq k \leq M\}$ and let $\{\sigma_k\}_{k=L}^{M}$ be an almost geometrically increasing sequence. Then*

$$\left\| \sigma_k \sum_{m=k}^{M} a_m \right\|_{\ell^q(S)} \approx \|\sigma_k a_k\|_{\ell^q(S)} \tag{1.3.5}$$

and

$$\|\sigma_k \sup_{k \leq m \leq M} a_m\|_{\ell^q(S)} \approx \|\sigma_k a_k\|_{\ell^q(S)} \tag{1.3.6}$$

for all non-negative sequences $\{a_k\}_{k=L}^{M}$.

1.4 Embeddings of sequence spaces

In Chapter 5 we shall make use of the following two assertions, which are well known. For the sake of completeness, we sketch their proofs.

Lemma 1.4.1 *Let $0 < P \le Q \le +\infty$, $\emptyset \ne \mathcal{Z} \subseteq \mathbb{Z}$ and let $\{w_k\}_{k\in\mathcal{Z}}$ and $\{v_k\}_{k\in\mathcal{Z}}$ be two sequences of positive numbers. Assume that*

$$\ell^P(\{v_k\},\mathcal{Z}) \hookrightarrow \ell^Q(\{w_k\},\mathcal{Z}). \tag{1.4.1}$$

Then

$$\|\{w_k v_k^{-1}\}\|_{\ell^\infty(\mathcal{Z})} \le C, \tag{1.4.2}$$

where C stands for the norm of the embedding (1.4.1).

Proof. Fix $n \in \mathcal{Z}$, test (1.4.1) with the sequence $\{a_k\}_{k\in\mathcal{Z}}$ given by

$$a_k := v_n^{-1}\chi_{\{n\}}(k), \quad k \in \mathcal{Z},$$

and then take the supremum of the result over all $n \in \mathcal{Z}$. □

Lemma 1.4.2 *Let $0 < Q < P \le +\infty$, $\emptyset \ne \mathcal{Z} \subseteq \mathbb{Z}$ and let $\{w_k\}_{k\in\mathcal{Z}}$ and $\{v_k\}_{k\in\mathcal{Z}}$ be two sequences of positive numbers. Assume that* (1.4.1) *holds. Then*

$$\|\{w_k v_k^{-1}\}\|_{\ell^R(\mathcal{Z})} \le C, \tag{1.4.3}$$

where $1/R := 1/Q - 1/P$ and C stands for the norm of the embedding (1.4.1).

Proof. Fix $n \in \mathbb{N}$, test (1.4.1) with the sequence $\{a_k\}_{k\in\mathcal{Z}}$ given by

$$a_k := \left(\frac{w_k}{v_k}\right)^{R/Q} \frac{1}{w_k}\,\chi_{[-n,n]}(k), \quad k \in \mathcal{Z},$$

and then take the limit of the result as $n \to +\infty$. □

1.5 Some reverse Hardy inequalities

The following two theorems concerning reverse Hardy inequalities will be needed in Chapter 5.

Theorem 1.5.1 (see [EGO1, Theorem 4.4]) *Assume that $I = (a,b) \subseteq \mathbb{R}$, $0 < p \le 1$, $p < q \le +\infty$, and $\mu,\nu \in \mathcal{B}^+(I)$. Let w and u be non-negative Borel measurable functions on I. Suppose that $\|u\|_{q,(a,t],\nu} < +\infty$ for all $t \in I$ and $u \ne 0$ ν-a.e. on I. Then the inequality*

$$\|g\|_{p,w,I,\mu} \le c\left\|\int_{(x,b)} g(y)\,d\mu\right\|_{q,u,I,\nu} \tag{1.5.1}$$

holds for all non-negative Borel measurable functions g on I, if and only if

$$C_1 := \left(\int_I \|w\|^r_{p^*,(a,t],\mu}\, d\left(-\|u\|^{-r}_{q,(a,t+],\nu}\right)\right)^{1/r} + \frac{\|w\|_{p^*,I,\mu}}{\|u\|_{q,I,\nu}} < +\infty,$$

where r is given by $1/r := 1/p - 1/q$ and $p^* := 1/(1-p)$.

Moreover, the best possible constant c in (1.5.1) *satisfies $c \approx C_1$.*

Theorem 1.5.2 (cf. [EGO1, Theorem 4.6 (ii)]) *Suppose that all the assumptions of* Theorem 1.5.1 *are satisfied. Let*

$$\|w\|_{p^*,(a,t],\mu} \lesssim \|w\|_{p^*,(a,t),\mu} \quad \text{for all} \quad t \in I. \tag{1.5.2}$$

Then $C_1 \approx C_2$, *where*

$$C_2 := \left(\int_I \|u\|_{q,(a,t+],\nu}^{-r} \, d\left(\|w\|_{p^*,(a,t+],\mu}^{r} \right) \right)^{1/r} + \frac{\lim\limits_{t\to a+} \|w\|_{p^*,(a,t),\mu}}{\lim\limits_{t\to a+} \|u\|_{q,(a,t),\nu}}.$$

Remark 1.5.3 *Let* μ *be a non-negative Borel measure on* I *which has no atoms. Then it is clear that the condition* (1.5.2) *holds.*

Suppose now that $p^* = +\infty$, *the function* w *satisfies* (1.5.2), *and* $C_2 < +\infty$. *Moreover, let* $\lim_{t\to a+} \|u\|_{q,(a,t),\nu} = 0$. *Then the second term in* C_2 *has to be finite which can happen only if* $\lim_{t\to a+} \|w\|_{\infty,(a,t),\mu} = 0$ (*cf. our* Convention 1.1.1 (i) *that* $0/0 = 0$).

Chapter 2

The set $Q_\rho(I)$ and its properties

2.1 Definitions and basic properties

Definition 2.1.1 Let $I = (a, b) \subseteq \mathbb{R}$ and $\rho \in Ads(I)$. We say that a function $h \in \mathcal{M}^+(I)$ is *ρ-quasiconcave on I* (and write $h \in Q_\rho(I)$) if

$$h \in \mathcal{M}^+(I;\uparrow) \quad \text{and} \quad \frac{h}{\rho} \in \mathcal{M}^+(I;\downarrow). \tag{2.1.1}$$

In the case that ρ is the identity map on I, that is, $\rho(t) = t$ for all $t \in I$, we write simply $Q(I)$ instead of $Q_\rho(I)$, and if $h \in Q(I)$, we say that h is *quasiconcave on I*.

If $h \in \mathcal{M}^+(I)$ and there is $g \in Q_\rho(I)$ (or $g \in Q(I)$) such that $h \approx g$ on I, we write $h \in Q_\rho^E(I)$ (or $h \in Q^E(I)$) and we say that h *is equivalent* to the ρ-quasiconcave (or quasiconcave) function g on I.

By $Q_\rho^0(I)$ we denote the subset of those $h \in Q_\rho(I)$ which satisfy

$$\lim_{t\to a+} h(t) = 0, \qquad \lim_{t\to b-} h(t) = +\infty, \tag{2.1.2}$$

$$\lim_{t\to a+} \left(\frac{\rho}{h}\right)(t) = 0, \qquad \lim_{t\to b-} \left(\frac{\rho}{h}\right)(t) = +\infty. \tag{2.1.3}$$

If ρ is the identity map on I, we write simply $Q^0(I)$ instead of $Q_\rho^0(I)$. The elements of $Q_\rho^0(I)$ (or $Q^0(I)$) are called *non-degenerate ρ-quasiconcave* (or *quasiconcave*) functions on I.

Remarks 2.1.2 (i) The following implications are true:

$$h \in Q_\rho(I),\ h \not\equiv 0 \Rightarrow h(x) \neq 0 \text{ for all } x \in I, \tag{2.1.4}$$

$$h \in Q_\rho(I),\ h \not\equiv +\infty \Rightarrow h(x) \neq +\infty \text{ for all } x \in I, \tag{2.1.5}$$

$$h \in Q_\rho(I) \Rightarrow h,\ \frac{h}{\rho} \in C(I). \tag{2.1.6}$$

The first two implications are trivial consequences of (2.1.1). To verify the last one, suppose the opposite. The points of discontinuity of $h \in \mathcal{M}^+(I;\uparrow)$ are the only points where h has jumps. Consider such a point, say x_0. Since $h \in \mathcal{M}^+(I;\uparrow)$, the values of h at points to the right of x_0 are greater than the values of h at those points which are to the left of x_0. The function $\frac{h}{\rho}$ has a jump at x_0 as well. Since any $\rho \in Ads(I)$ is continuous, it follows from what we said above that in some neighbourhood of x_0 there are points close to the right of x_0, where $\frac{h}{\rho}$ has greater values than at points which are to the left of x_0. But this contradicts the inclusion $\frac{h}{\rho} \in \mathcal{M}^+(I;\downarrow)$.

(ii) Let $\rho \in Ads(I)$, $h \in \mathcal{M}^+(I)$, and $x \in I$. Then $h \in Q_\rho(I)$, if and only if

$$h(t) \leq \max\left\{1, \frac{\rho(t)}{\rho(x)}\right\} h(x) \quad \text{for all } t \in I. \tag{2.1.7}$$

Indeed, let $h \in Q_\rho(I)$ and $x \in I$. If $t \in I$ and $t \leq x$, then the fact that $h, \rho \in \mathcal{M}^+(I;\uparrow)$ implies that $h(t) \leq h(x) = \max\{1, \rho(t)/\rho(x)\}h(x)$. When $t \in I$ and $t > x$, then the inclusion $\frac{h}{\rho} \in \mathcal{M}^+(I;\downarrow)$, yields $h(t)/\rho(t) \leq h(x)/\rho(x)$. Since $\rho \in \mathcal{M}^+(I;\uparrow)$, the last inequality implies that $h(t) \leq \rho(t)h(x)/\rho(x) = \max\{1, \rho(t)/\rho(x)\}h(x)$ and (2.1.7) is proved. To verify the opposite implication, assume now that $\rho \in Ads(I)$, $h \in \mathcal{M}^+(I)$, $x \in I$ and that (2.1.7) holds. If $t \in I$ and $t \leq x$, then $\rho(t)/\rho(x) \leq 1$. Thus, (2.1.7) implies that $h(t) \leq h(x)$. Consequently, $h \in \mathcal{M}^+(I;\uparrow)$. Moreover, if $t \in I$ and $t > x$, then $\rho(t)/\rho(x) \geq 1$. Thus, (2.1.7) implies that $h(t) \leq \rho(t)h(x)/\rho(x)$, i.e., $h(t)/\rho(t) \leq h(x)/\rho(x)$. Consequently, $\frac{h}{\rho} \in \mathcal{M}^+(I;\downarrow)$, and the proof is complete.

Note that if $I = (0, +\infty)$ and ρ is the identity map on I, then in [BK], p. 291, any non-negative continuous function h on I satisfying inequality (2.1.7) is called a quasiconcave function. It follows from (2.1.6) that in such a definition the assumption that h is continuous is redundant.

In the following lemma we list some basic properties of the class $Q_\rho(I)$.

Lemma 2.1.3 *Let $I = (a, b) \subseteq \mathbb{R}$.*

(i) If $\rho \in Ads(I)$, $h, h_1, h_2 \in Q_\rho(I)$, $c_1, c_2 \in [0, +\infty)$ and $p \in (0, +\infty)$, then

$$c_1 h_1 + c_2 h_2 \in Q_\rho(I), \tag{2.1.8}$$

$$h^p \in Q_{\rho^p}(I), \tag{2.1.9}$$

$$\frac{\rho}{h} \in Q_\rho(I). \tag{2.1.10}$$

Moreover, if f is an increasing[1] *continuous function which maps the interval I onto the interval $I_1 := f(I)$, then*

$$h \circ f^{-1} \in Q_{\rho \circ f^{-1}}(I_1). \tag{2.1.11}$$

Similarly, if f is a decreasing continuous function, $I_1 := f(I)$ and $\widetilde{\rho} = 1/\rho \circ f^{-1}$, then

$$\widetilde{\rho} \cdot (h \circ f^{-1}) \in Q_{\widetilde{\rho}}(I_1). \tag{2.1.12}$$

[1] By increasing (decreasing) we mean strictly increasing (decreasing).

(ii) *If* $\rho_1, \rho_2 \in Ads(I)$, $h_1 \in Q_{\rho_1}(I)$, $h_2 \in Q_{\rho_2}(I)$, *then*

$$h_1 h_2 \in Q_{\rho_1\rho_2}(I). \tag{2.1.13}$$

Proof. This is evident and so left to the reader. □

2.2 Examples of ρ-quasiconcave functions

Further properties and examples of ρ-quasiconcave functions we begin with some simple examples.

Examples 2.2.1 Assume that $I = (a,b) \subseteq \mathbb{R}$ and $\rho \in Ads(I)$.

(i) Let $x \in I$ and $c \in (0,+\infty)$. Then the functions given by

$$h_1(t) := c \quad \text{for all } t \in I,$$

$$h_2(t) := \rho(t) \quad \text{for all } t \in I,$$

$$h_3(t) := \min\{\rho(t), \rho(x)\} \quad \text{for all } t \in I, \tag{2.2.1}$$

$$h_4(t) := \max\{\rho(t), \rho(x)\} \quad \text{for all } t \in I, \tag{2.2.2}$$

$$h_5(t) := \min\left\{\frac{\rho(t)}{\rho(x)}, 1\right\} \quad \text{for all} \quad t \in I, \tag{2.2.3}$$

$$h_6(t) := \max\left\{\frac{\rho(t)}{\rho(x)}, 1\right\} \quad \text{for all} \quad t \in I$$

are ρ-quasiconcave on I.

(ii) Let $\mu \in \mathcal{B}^+(I)$, $w \in \mathcal{M}^+(I,\mu)$ and $p \in (0,+\infty]$. Then the function h_7 defined by

$$h_7(x) := \|\min\{\rho(\cdot), \rho(x)\}\|_{p,w,I,\mu}, \quad x \in I, \tag{2.2.4}$$

is ρ-quasiconcave on I. Indeed, putting

$$\nu(t,x) := \min\{\rho(t), \rho(x)\} \quad \text{and} \quad \widetilde{\nu}(t,x) := \nu(t,x)/\rho(x) \quad \text{for } t, x \in I,$$

we see that

$$\nu(t,x) = \begin{cases} \rho(t), & t \in (a,x] \\ \rho(x), & t \in (x,b) \end{cases} \quad \text{and} \quad \widetilde{\nu}(t,x) = \begin{cases} \rho(t)/\rho(x), & t \in (a,x] \\ 1, & t \in (x,b). \end{cases} \tag{2.2.5}$$

Let $x_1, x_2 \in I$, $x_1 < x_2$. Then (2.2.5) implies that

$$\nu(t,x_1) \le \nu(t,x_2) \quad \text{and} \quad \widetilde{\nu}(t,x_1) \ge \widetilde{\nu}(t,x_2) \quad \text{for all } t \in I.$$

Since the functional $\|\cdot\|_{p,w,I,\mu}$ is monotone in the sense that $\|g\|_{p,w,I,\mu} \le \|f\|_{p,w,I,\mu}$ if $g, f \in \mathcal{M}^+(I,\mu)$ and $g \le f$ μ-a.e. on I, we obtain that

$$h_7(x_1) = \|\nu(\cdot,x_1)\|_{p,w,I,\mu} \le \|\nu(\cdot,x_2)\|_{p,w,I,\mu} = h_7(x_2)$$

and

$$\left(\frac{h_7}{\rho}\right)(x_1) = \|\widetilde{\nu}(\cdot, x_1)\|_{p,w,I,\mu} \geq \|\widetilde{\nu}(\cdot, x_2)\|_{p,w,I,\mu} = \left(\frac{h_7}{\rho}\right)(x_2),$$

and the result follows.

Similarly, one can verify that the function h_8 defined by

$$h_8(x) := \|\max\{\rho(\cdot), \rho(x)\}\|_{p,w,I,\mu}, \quad x \in I, \tag{2.2.6}$$

is ρ-quasiconcave on I.

(iii) Let $\mu \in \mathcal{B}^+(I)$, $w \in \mathcal{M}^+(I,\mu)$ and $p \in (0,+\infty]$. Then the inclusion $h_7 \in Q_\rho(I)$ implies that the functions

$$h_9(x) := \left\|\frac{\rho(x)}{\rho(\cdot)+\rho(x)}\right\|_{p,w,I,\mu}, \quad x \in I, \tag{2.2.7}$$

$$h_{10}(x) := \left\|\frac{\rho(\cdot)\rho(x)}{\rho(\cdot)+\rho(x)}\right\|_{p,w,I,\mu}, \quad x \in I, \tag{2.2.8}$$

belong to $Q_\rho^E(I)$. Indeed, we have

$$\begin{aligned} h_9(x) &= \left\|\frac{1}{\frac{\rho(\cdot)}{\rho(x)}+1}\right\|_{p,w,I,\mu} \approx \left\|\frac{1}{\max\{\frac{\rho(\cdot)}{\rho(x)},1\}}\right\|_{p,w,I,\mu} \\ &= \left\|\min\left\{\frac{\rho(x)}{\rho(\cdot)},1\right\}\right\|_{p,w,I,\mu} = \|\min\{\rho(x),\rho(\cdot)\}\|_{p,w/\rho,I,\mu}, \ x \in I, \end{aligned}$$

and so the inclusion $h_9 \in Q_\rho^E(I)$ is a consequence of the fact that $h_7 \in Q_\rho(I)$. Similarly,

$$\begin{aligned} h_{10}(x) &= \left\|\frac{1}{\frac{1}{\rho(x)}+\frac{1}{\rho(\cdot)}}\right\|_{p,w,I,\mu} \approx \left\|\frac{1}{\max\{\frac{1}{\rho(x)},\frac{1}{\rho(\cdot)}\}}\right\|_{p,w,I,\mu} \\ &= \|\min\{\rho(x),\rho(\cdot)\}\|_{p,w,I,\mu}, \quad x \in I. \end{aligned}$$

Therefore, the conclusion $h_{10} \in Q_\rho^E(I)$ also follows from the fact that $h_7 \in Q_\rho(I)$.

On the other hand, it is even possible to show that

$$h_9, h_{10} \in Q_\rho(I).$$

To verify that $h_9 \in Q_\rho(I)$, fix $t \in I$ and put $f(y) := y/(\rho(t)+y)$ for $y \in J := (0,+\infty)$. Since $f' > 0$ on J, the function f is non-decreasing on J. Together with the fact that the function ρ is non-decreasing on I, this implies that

$$\frac{\rho(x_1)}{\rho(t)+\rho(x_1)} = f(\rho(x_1)) \leq f(\rho(x_2)) = \frac{\rho(x_2)}{\rho(t)+\rho(x_2)} \text{ if } x_1, x_2 \in I \text{ and } x_1 \leq x_2.$$

Therefore, if $x_1, x_2 \in I$ and $x_1 \leq x_2$, then

$$h_9(x_1) = \left\|\frac{\rho(x_1)}{\rho(\cdot)+\rho(x_1)}\right\|_{p,w,I,\mu} \leq \left\|\frac{\rho(x_2)}{\rho(\cdot)+\rho(x_2)}\right\|_{p,w,I,\mu} = h_9(x_2),$$

i.e., the function h_9 is non-decreasing on the interval I. On putting $g(y) := 1/(\rho(t)+y)$ for $y \in J := (0,+\infty)$, we see that the function g is non-increasing on J. Hence,

$$\frac{1}{\rho(t)+\rho(x_1)} = g(\rho(x_1)) \geq g(\rho(x_2)) = \frac{1}{\rho(t)+\rho(x_2)} \text{ if } x_1, x_2 \in I \text{ and } x_1 \leq x_2.$$

Therefore, if $x_1, x_2 \in I$ and $x_1 \leq x_2$, then

$$\frac{h_9(x_1)}{\rho(x_1)} = \left\| \frac{1}{\rho(\cdot)+\rho(x_1)} \right\|_{p,w,I,\mu} \geq \left\| \frac{1}{\rho(\cdot)+\rho(x_2)} \right\|_{p,w,I,\mu} = \frac{h_9(x_2)}{\rho(x_2)},$$

i.e., the function h_9/ρ is non-decreasing on the interval I, and the proof of the inclusion $h_9 \in Q_\rho(I)$ is complete.

The proof that $h_{10} \in Q_\rho(I)$ is analogous (now, for a fix $t \in I$, one uses the functions $f(y) := (\rho(t)\, y)/(\rho(t)+y)$ and $g(y) := \rho(t)/(\rho(t)+y)$ for $y \in J := (0,+\infty)$).

Remarks 2.2.2 (i) Let $I = (a,b) \subseteq \mathbb{R}$. Recall that the function $f : I \to \mathbb{R}$ is said to be *concave* on I if

$$f(\lambda x + (1-\lambda)y) \geq \lambda f(x) + (1-\lambda) f(y) \tag{2.2.9}$$

for all $x, y \in I$ and every $\lambda \in (0,1)$. (Condition (2.2.9) means geometrically that the point $[t, f(t)]$ lies, for t between x and y, above the line segment joining the points $[x, f(x)]$ and $[y, f(y)]$.) Recall also that a concave function on I is continuous on I.

Let f be non-negative and concave on I. Then the following implications hold:

$$b = +\infty \Rightarrow f \in \mathcal{M}^+(I;\uparrow), \tag{2.2.10}$$

$$a = 0 \Rightarrow \frac{f}{Id} \in \mathcal{M}^+(I;\downarrow).\text{[2]} \tag{2.2.11}$$

To verify (2.2.10), let $s \in I$ and $t > s$. Choose T such that $s < t < T$ and put $\lambda = (t-s)/(T-s)$. Then

$$f(t) = f((1-\lambda)s + \lambda T) \geq (1-\lambda) f(s) + \lambda f(T) \geq (1-\lambda) f(s) = \frac{T-t}{T-s} f(s),$$

which implies that

$$f(t) \geq \lim_{T \to +\infty} \frac{T-t}{T-s} f(s) = f(s),$$

and (2.2.10) follows.

To verify (2.2.11), let $s, t \in I$ and $s < t$. Choose r such that $0 < r < s < t$ and put $\lambda = (s-r)/(t-r)$. Then

$$f(s) = f((1-\lambda)r + \lambda t) \geq (1-\lambda) f(r) + \lambda f(t) \geq \lambda f(t) = \frac{s-r}{t-r} f(t),$$

[2] By Id we mean the identity map on I.

which implies that

$$f(s) \geq \lim_{r \to 0+} \frac{s-r}{t-r} f(t) = \frac{s}{t} f(t),$$

and (2.2.11) follows.

By (2.2.10) and (2.2.11), any non-negative concave function on $(0, +\infty)$ is quasiconcave on $(0, +\infty)$. However, if the interval $(0, +\infty)$ is replaced by a general interval $(a, b) \subseteq \mathbb{R}$ such an assertion does not hold. To see this, let $h(t) = t - a$ for $t \in (a, b)$, where $0 < a < b \leq +\infty$. Then h is non-negative and concave on (a, b), but not quasiconcave since the function $\frac{h(t)}{t} = 1 - \frac{a}{t}$, $t \in (a, b)$, is increasing on (a, b).

(ii) If $h \in Q(I)$, $I \subseteq (0, +\infty)$, and $x \in I$, then it follows from (2.1.7) that a function h is dominated on I by the concave function

$$\psi(t) := \Big(1 + \frac{t}{x}\Big) h(x), \quad t \in I,$$

that is, $h \leq \psi$ on I. Since the pointwise infimum of concave functions is itself concave, we observe that there is the smallest concave function on I, say $\widetilde{h}$, which dominates h on I. Such a function is called *the least concave majorant* of h on I. As ψ is concave on I, we have $\widetilde{h}(x) \leq \psi(x) = 2h(x)$. Consequently, the least concave majorant $\widetilde{h}$ of $h \in Q(I)$ satisfies

$$h \leq \widetilde{h} \leq 2h \quad \text{on} \quad I. \tag{2.2.12}$$

(iii) Let $I = (a, b) \subseteq \mathbb{R}$ and let $f : I \to \mathbb{R}$ be a function on I. Then (cf., e.g., [RR, Chap. I, Th. 1]), f is concave on I, if and only if, for each closed interval $[c, d] \subset I$,

$$f(x) = f(c) + \int_c^x \theta(y)\, dy, \quad x \in [c, d], \tag{2.2.13}$$

where $\theta : I \to \mathbb{R}$ is a non-increasing and right continuous function on I. (Note that θ is the derivative of f a.e. on I.)

To present some more examples of ρ-quasiconcave functions, we shall use the following lemma.

Lemma 2.2.3 *Let $I = (a, b) \subseteq \mathbb{R}$ and $G \in \mathcal{M}^+(I; \downarrow)$. Let $u \in \mathcal{W}(I)$ be such that the function ρ given by*

$$\rho(x) := \int_a^x u(t)\, dt, \quad x \in I, \tag{2.2.14}$$

satisfies $\rho \in Ads(I)$. Put

$$F(x) := \frac{\int_a^x G(t) u(t)\, dt}{\rho(x)}, \quad x \in I. \tag{2.2.15}$$

Then

$$F \in \mathcal{M}^+(I;\downarrow) \tag{2.2.16}$$

and

$$F\rho \in Q_\rho(I). \tag{2.2.17}$$

Proof. It is clear that the function $H := G \circ \rho^{-1}$ satisfies

$$H \in \mathcal{M}^+(J;\downarrow), \quad \text{where} \quad J = \rho(I) = (\rho(a+), \rho(b-)). \tag{2.2.18}$$

Moreover,

$$F(z) = \frac{\int_a^z H(\rho(t))\, d\rho(t)}{\rho(z)} \quad \text{for all } z \in I.$$

Using the change of variables $s = \rho(t)$ in the last integral and the fact that $\rho(a+) = 0$, we arrive at

$$F(z) = \frac{\int_0^{\rho(z)} H(s)\, ds}{\rho(z)} \quad \text{for all } z \in I. \tag{2.2.19}$$

Now, let $x, y \in I$ and $x < y$. Then (2.2.19) and the change of variables $s = t\rho(x)/\rho(y)$ imply that

$$F(x) = \frac{\int_0^{\rho(x)} H(s)\, ds}{\rho(x)} = \frac{\int_0^{\rho(y)} H(t\frac{\rho(x)}{\rho(y)})\, dt}{\rho(y)}. \tag{2.2.20}$$

Since $t\rho(x)/\rho(y) \le t$ for all $t \ge 0$, we obtain from (2.2.18) and (2.2.19) that

$$\text{RHS}(2.2.20) \ge \frac{\int_0^{\rho(y)} H(t)\, dt}{\rho(y)} = F(y).$$

Together with (2.2.20), this yields (2.2.16).

Since, by (2.2.15), $F(x)\rho(x) = \int_a^x G(t)u(t)\, dt$, $x \in I$, it is clear that $F\rho \in \mathcal{M}^+(I;\uparrow)$. This relation and (2.2.16) yield (2.2.17). □

Corollary 2.2.4 *Suppose that $I = (a,b) \subseteq \mathbb{R}$ and that $u \in \mathcal{W}(I)$ is such that the function ρ given by $\rho(x) := \int_a^x u(t)\, dt$, $x \in I$, satisfies $\rho \in Ads(I)$. Let $g \in \mathcal{M}^+(I)$. Then the functions*

$$h_{11}(x) := \int_a^x u(t) \Big(\int_t^b g(s)\, ds \Big)\, dt, \quad x \in I, \tag{2.2.21}$$

and

$$h_{12}(x) := \int_a^x u(t)\, \|g\|_{\infty,(t,b)}\, dt, \quad x \in I, \tag{2.2.22}$$

belong to $Q_\rho(I)$.

Proof. As the functions $G_{10}(x) := \int_x^b g(s)\,ds$ and $G_{11}(x) := \|g\|_{\infty,(x,b)}$, $x \in I$, belong to $\mathcal{M}^+(I, \downarrow)$, the result follows from Lemma 2.2.3. □

Remark 2.2.5 It is instructive to give another proof of the relation $h_{11} \in Q_\rho(I)$. Applying Fubini's theorem, we obtain

$$\begin{aligned} h_{11}(x) &= \int_a^x \Big(\int_a^s u(t)\,dt\Big) g(s)\,ds + \Big(\int_a^x u(t)\,dt\Big) \int_x^b g(s)\,ds \\ &= \int_a^b \min\{\rho(s), \rho(x)\} g(s)\,ds \\ &= \|\min\{\rho(\cdot), \rho(x)\}\|_{1,g,I}, \quad x \in I. \end{aligned}$$

Consequently, the result follows from the fact that the function h_7 from (2.2.4) is ρ-quasiconcave on I.

The following is a counterpart of Lemma 2.2.3.

Lemma 2.2.6 *Let $I = (a,b) \subseteq \mathbb{R}$ and $G \in \mathcal{M}^+(I;\uparrow)$. Let $u \in \mathcal{W}(I)$ be such that the function $\widetilde{\rho}$ given by*

$$\widetilde{\rho}(x) := \Big(\int_x^b u(t)\,dt\Big)^{-1}, \quad x \in I \tag{2.2.23}$$

satisfies $\widetilde{\rho} \in Ads(I)$. Put

$$F(x) := \frac{\int_x^b G(t)u(t)\,dt}{\int_x^b u(t)\,dt}, \quad x \in I. \tag{2.2.24}$$

Then

$$F \in \mathcal{M}^+(I;\uparrow) \tag{2.2.25}$$

and

$$F \in Q_{\widetilde{\rho}}(I). \tag{2.2.26}$$

Proof. Putting $\Phi(x) := -\int_x^b u(t)\,dt$, $x \in I$, we see that the function $H := G \circ \Phi^{-1}$ satisfies

$$H \in \mathcal{M}^+(J;\uparrow), \quad \text{where} \quad J = \Phi(I) = (\Phi(a+), \Phi(b-)). \tag{2.2.27}$$

Moreover,

$$F(z) = \frac{\int_z^b H(\Phi(t))\,d\Phi(t)}{-\Phi(z)} \quad \text{for all} \quad z \in I.$$

Using the change of variables $s = \Phi(t)$ in the last integral and the fact that $\Phi(b-) = 0$, we arrive at

$$F(z) = \frac{\int_{\Phi(z)}^0 H(s)\,ds}{-\Phi(z)} \quad \text{for all} \quad z \in I. \tag{2.2.28}$$

Now, let $x, y \in I$ and $x < y$. Then (2.2.28) and the change of variables $s = t\Phi(x)/\Phi(y)$ imply that

$$F(x) = \frac{\int_{\Phi(x)}^{0} H(s)\,ds}{-\Phi(x)} = \frac{\int_{\Phi(y)}^{0} H(t\frac{\Phi(x)}{\Phi(y)})\,dt}{-\Phi(y)}. \tag{2.2.29}$$

Since $t\Phi(x)/\Phi(y) \le t$ for all $t \le 0$, we obtain from (2.2.27) and (2.2.28) that

$$\text{RHS}(2.2.29) \le \frac{\int_{\Phi(y)}^{0} H(t)\,dt}{-\Phi(y)} = F(y).$$

Together with (2.2.29), this yields (2.2.25).

Since, by (2.2.23) and (2.2.24), $F(x)/\widetilde{\rho}(x) = \int_x^b G(t)u(t)\,dt$, $x \in I$, it is clear that $F/\widetilde{\rho} \in \mathcal{M}^+(I;\downarrow)$. This relation and (2.2.25) yield (2.2.26). □

Corollary 2.2.7 *Suppose that $I = (a, b) \subseteq \mathbb{R}$ and that $u \in \mathcal{W}(I)$ is such that the function $\widetilde{\rho}$ given by $\widetilde{\rho}(x) := (\int_x^b u(t)\,dt)^{-1}$, $x \in I$, satisfies $\widetilde{\rho} \in Ads(I)$. Let $g \in \mathcal{M}^+(I)$. Then the functions*

$$h_{13}(x) := \widetilde{\rho}(x) \int_x^b u(t) \Big(\int_a^t g(s)\,ds \Big)\,dt, \quad x \in I, \tag{2.2.30}$$

and

$$h_{14}(x) := \widetilde{\rho}(x) \int_x^b u(t)\, \|g\|_{\infty,(a,t)}\,dt, \quad x \in I, \tag{2.2.31}$$

belong to $Q_{\widetilde{\rho}}(I)$.

Proof. As the functions $G_{12}(x) := \int_a^x g(s)\,ds$ and $G_{13}(x) := \|g\|_{\infty,(a,x)}$, $x \in I$, belong to $\mathcal{M}^+(I;\uparrow)$, the result follows from Lemma 2.2.6. □

Remarks 2.2.8 (i) There are other proofs of the relation $h_{13} \in Q_{\widetilde{\rho}}(I)$.

(*i*-1) Using Fubini's theorem, we obtain

$$\begin{aligned}\frac{h_{13}(x)}{\widetilde{\rho}(x)} &= \Big(\int_a^x g(s)\,ds \Big) \Big(\int_x^b u(t)\,dt \Big) + \int_x^b g(s) \Big(\int_s^b u(t)\,dt \Big)\,ds \\ &= \frac{1}{\widetilde{\rho}(x)} \int_a^x g(s)\,ds + \int_x^b \frac{1}{\widetilde{\rho}(s)} g(s)\,ds, \quad x \in I,\end{aligned}$$

which implies that

$$\begin{aligned}h_{13}(x) &= \int_a^x g(s)\,ds + \int_x^b \frac{\widetilde{\rho}(x)}{\widetilde{\rho}(s)} g(s)\,ds \\ &= \int_a^b \min\Big\{1, \frac{\widetilde{\rho}(x)}{\widetilde{\rho}(s)}\Big\} g(s)\,ds \\ &= \|\min\{\widetilde{\rho}(\cdot), \widetilde{\rho}(x)\}\|_{1,g/\widetilde{\rho},I}, \ x \in I.\end{aligned}$$

Consequently, the result follows from the fact that the function h_7 from (2.2.4) is ρ-quasiconcave on I.

(*i*-2) Put $\widetilde{I} = (-b, -a)$ and define the functions $\rho, \widetilde{u}$ and $\widetilde{g}$ on $\widetilde{I}$ by

$$\rho(x) = \frac{1}{\widetilde{\rho}(-x)}, \quad \widetilde{u}(x) = u(-x), \quad \widetilde{g}(x) = g(-x). \tag{2.2.32}$$

Then $\rho \in Ads(\widetilde{I})$, $\widetilde{u} \in \mathcal{W}(\widetilde{I})$ and $\widetilde{g} \in \mathcal{M}^+(\widetilde{I})$. Moreover, for all $x \in \widetilde{I}$,

$$\frac{h_{13}(-x)}{\widetilde{\rho}(-x)} = \int_{-x}^{b} u(t) \left(\int_a^t g(s)\, ds \right) dt, \tag{2.2.33}$$

and, using (2.2.32) and the change of variables $t = -\widetilde{t}$, $s = -\widetilde{s}$, we arrive at

$$\frac{h_{13}(-x)}{\widetilde{\rho}(-x)} = \int_{-b}^{x} \widetilde{u}(\widetilde{t}) \left(\int_{\widetilde{t}}^{-a} \widetilde{g}(\widetilde{s})\, d\widetilde{s} \right) d\widetilde{t} =: \widetilde{h}(x) \text{ for all } x \in \widetilde{I}. \tag{2.2.34}$$

Furthermore, using (2.2.32), (2.2.23) and the change of variables $t = -\widetilde{t}$, we obtain

$$\rho(x) = \frac{1}{\widetilde{\rho}(-x)} = \int_{-x}^{b} u(t)\, dt = \int_{-b}^{x} \widetilde{u}(\widetilde{t})\, d\widetilde{t} \text{ for all } x \in \widetilde{I}. \tag{2.2.35}$$

Thus, we have by (2.2.34), (2.2.35) and Corollary 2.2.4 that

$$\widetilde{h} \in Q_\rho(\widetilde{I}). \tag{2.2.36}$$

Now, putting $f(x) = -x$, $x \in \widetilde{I}$, and observing that

$$\widetilde{\rho}(x) = \frac{1}{\rho(-x)} = \frac{1}{(\rho \circ f^{-1})(x)} \quad \text{if} \quad x \in I, \tag{2.2.37}$$

we have (cf. (2.1.12))

$$\widetilde{\rho} \cdot (\widetilde{h} \circ f^{-1}) \in Q_{\widetilde{\rho}}(I).$$

Since, by (2.2.34), $\widetilde{\rho} \cdot (\widetilde{h} \circ f^{-1}) = h_{13}$, the proof is complete.

(ii) The "symmetry" argument of part (*i*-2) and the fact that the function h_{12} from (2.2.22) is ρ-quasiconcave provide another proof of the inclusion $h_{14} \in Q_{\widetilde{\rho}}(I)$.

There is the following variant of Lemma 2.2.3.

Lemma 2.2.9 *Let $I = (a, b) \subseteq \mathbb{R}$, $\rho \in Ads(I)$ and let $G \in \mathcal{M}^+(I; \downarrow)$ be a right continuous function on I. Put*

$$F(x) := \frac{\|G\rho\|_{\infty,(a,x)}}{\rho(x)}, \quad x \in I. \tag{2.2.38}$$

Then

$$F\rho \in Q_\rho(I). \tag{2.2.39}$$

Proof. The assumptions on G imply that

$$G(x) = \|G\|_{\infty,(x,b)} \quad \text{for all } x \in I.$$

Moreover (cf. [OK, Lemma 5.3]),

$$G(x)\rho(x) \leq \|G\rho\|_{\infty,(a,x)} \quad \text{for a.e. } x \in I.$$

Therefore, for a.e. $x \in I$,

$$\begin{aligned} F(x)\rho(x) &= \|G\rho\|_{\infty,(a,x)} = \max\{\|G\rho\|_{\infty,(a,x)}, G(x)\rho(x)\} \\ &= \max\{\|G\rho\|_{\infty,(a,x)},\ \rho(x)\|G\|_{\infty,(x,b)}\} \\ &= \|\min\{\rho(\cdot), \rho(x)\}\|_{\infty,G,I}. \end{aligned}$$

Since, by Example 2.2.1 (ii), the function $h(x) := \|\min\{\rho(\cdot), \rho(x)\}\|_{\infty,G,I}$, $x \in I$, belongs to $Q_\rho(I)$, it is sufficient to prove that

$$S := F\rho \in C(I). \tag{2.2.40}$$

The function

$$S(x) = \|G\rho\|_{\infty,(a,x)}, \quad x \in I, \tag{2.2.41}$$

is left continuous on I and $S \in \mathcal{M}^+(I;\uparrow)$. Consequently, if $\bar{x} \in I$ and S is not continuous at $\bar{x}$, then

$$S(\bar{x}) < L := \lim_{x \to \bar{x}+} S(x).$$

Together with (2.2.41), this implies that there are two monotone sequences $\{t_n\}_{n\in\mathbb{N}}$, $t_n \leq t_{n+1} \leq \bar{x}$ for $n \in \mathbb{N}$, and $\{\tau_n\}_{n\in\mathbb{N}}$, $\bar{x} \leq \tau_{n+1} \leq \tau_n$ for $n \in \mathbb{N}$, such that $\lim_{n\to\infty} t_n = \bar{x} = \lim_{n\to\infty} \tau_n$ and

$$G(t_n)\rho(t_n) \leq S(\bar{x}) < L \leq G(\tau_n)\rho(\tau_n) \quad \text{for all } n \in \mathbb{N}.$$

Hence,

$$\rho(\bar{x}) \lim_{n\to+\infty} G(t_n) \leq S(\bar{x}) < L \leq \rho(\bar{x})G(\bar{x}),$$

which yields

$$G(t_n) < \frac{L}{\rho(\bar{x})} \leq G(\bar{x}) \quad \text{for all large enough } n \in \mathbb{N}.$$

But this is impossible since $G \in \mathcal{M}^+(I;\downarrow)$. Therefore (2.2.40) holds. □

Corollary 2.2.10 *Let $I = (a,b) \subseteq \mathbb{R}$, $\rho \in Ads(I)$ and $g \in \mathcal{M}^+(I)$. Then the functions*

$$h_{15}(x) := \left\|\rho(t)\int_t^b g(s)\,ds\right\|_{\infty,(a,x)}, \quad x \in I, \tag{2.2.42}$$

and

$$h_{16}(x) := \|\rho(t)\ \|g\|_{\infty,(t,b)}\|_{\infty,(a,x)}, \quad x \in I, \tag{2.2.43}$$

belong to $Q_\rho(I)$.

Proof. As the functions $G_{14}(x) := \int_x^b g(s)\,ds$ and $G_{15}(x) := \|g\|_{\infty,(x,b)}$, $x \in I$, are right continuous on I and belong to $\mathcal{M}^+(I;\downarrow)$, the result follows from Lemma 2.2.9.

□

A symmetric counterpart of Lemma 2.2.9 reads as follows.

Lemma 2.2.11 *Let $I = (a,b) \subseteq \mathbb{R}$, $\widetilde{\rho} \in Ads(I)$ and let $G \in \mathcal{M}^+(I;\uparrow)$ be a left continuous function on I. Put*

$$F(x) := \widetilde{\rho}(x)\|G/\widetilde{\rho}\|_{\infty,(x,b)}, \quad x \in I. \tag{2.2.44}$$

Then

$$F \in Q_{\widetilde{\rho}}(I). \tag{2.2.45}$$

Proof. Put $\widetilde{I} = (-b,-a)$ and define the functions ρ and $\widetilde{G}$ on $\widetilde{I}$ by

$$\rho(x) = \frac{1}{\widetilde{\rho}(-x)}, \quad \widetilde{G}(x) = G(-x). \tag{2.2.46}$$

Then $\rho \in Ads(I)$. Moreover, for all $x \in \widetilde{I}$,

$$\frac{F(-x)}{\widetilde{\rho}(-x)} = \|G(t)/\widetilde{\rho}(t)\|_{\infty,(-x,b)}, \tag{2.2.47}$$

and, using (2.2.46) and the change of variables $t = -\widetilde{t}$, we arrive at

$$\frac{F(-x)}{\widetilde{\rho}(-x)} = \|\widetilde{G}(\widetilde{t})\rho(\widetilde{t})\|_{\infty,(-b,x)} =: \widetilde{h}(x) \text{ for all } x \in \widetilde{I}. \tag{2.2.48}$$

Thus, by Lemma 2.2.9,

$$\widetilde{h} \in Q_\rho(\widetilde{I}).$$

Now, putting $f(x) = -x$, $x \in \widetilde{I}$, and observing that (2.2.37) is satisfied, we have (cf. (2.1.12))

$$\widetilde{\rho}\cdot(\widetilde{h}\circ f^{-1}) \in Q_{\widetilde{\rho}}(I).$$

Since, by (2.2.48), $\widetilde{\rho}\cdot(\widetilde{h}\circ f^{-1}) = F$, the proof is complete. □

Corollary 2.2.12 *Let $I = (a,b) \subseteq \mathbb{R}$, $\widetilde{\rho} \in Ads(I)$ and $g \in \mathcal{M}^+(I)$. Then the functions*

$$h_{17}(x) := \widetilde{\rho}(x)\left\|\frac{1}{\widetilde{\rho}(t)}\int_a^t g(s)\,ds\right\|_{\infty,(x,b)}, \quad x \in I, \tag{2.2.49}$$

and

$$h_{18}(x) := \widetilde{\rho}(x)\left\|\frac{1}{\widetilde{\rho}(t)}\|g\|_{\infty,(a,t)}\right\|_{\infty,(x,b)}, \quad x \in I, \tag{2.2.50}$$

belong to $Q_{\widetilde{\rho}}(I)$.

Remarks 2.2.13 (i) There are also alternative proofs of the relations $h_{16} \in Q_\rho(I)$ and $h_{18} \in Q_{\widetilde{\rho}}(I)$. Indeed, exchanging the order of the essential suprema on the right-hand side of (2.2.43), we obtain

$$\begin{aligned}(2.2.51)\quad h_{16}(x) &= \max\{\|g(s)\ \|\rho(t)\|_{\infty,(a,s)}\|_{\infty,(a,x)}, \|g(s)\ \|\rho(t)\|_{\infty,(a,x)}\|_{\infty,(x,b)}\}\\ &= \max\{\|g(s)\rho(s)\|_{\infty,(a,x)}, \rho(x)\ \|g(s)\|_{\infty,(x,b)}\}\\ &= \|\min\{\rho(\cdot),\rho(x)\}\|_{\infty,g,I} \quad \text{for all } x \in I.\end{aligned}$$

Thus, the relation $h_{16} \in Q_\rho(I)$ follows from Example 2.2.1 (ii).

Similarly, we obtain

$$\begin{aligned}\frac{h_{18}(x)}{\widetilde{\rho}(x)} &= \max\left\{\left\|g(s)\left\|\frac{1}{\widetilde{\rho}(t)}\right\|_{\infty,(x,b)}\right\|_{\infty,(a,x)}, \left\|g(s)\left\|\frac{1}{\widetilde{\rho}(t)}\right\|_{\infty,(s,b)}\right\|_{\infty,(x,b)}\right\}\\ &= \max\left\{\frac{1}{\widetilde{\rho}(x)}\|g(s)\|_{\infty,(a,x)}, \left\|g(s)\frac{1}{\widetilde{\rho}(s)}\right\|_{\infty,(x,b)}\right\}\\ &= \frac{1}{\widetilde{\rho}(x)}\max\left\{\|g(s)\|_{\infty,(a,x)}, \left\|g(s)\frac{\widetilde{\rho}(x)}{\widetilde{\rho}(s)}\right\|_{\infty,(x,b)}\right\}\\ &= \frac{1}{\widetilde{\rho}(x)}\left\|\min\left\{1,\frac{\widetilde{\rho}(x)}{\widetilde{\rho}(s)}\right\}g(s)\right\|_{\infty,I}\\ &= \frac{1}{\widetilde{\rho}(x)}\|\min\{\widetilde{\rho}(\cdot),\widetilde{\rho}(x)\}\|_{\infty,g/\widetilde{\rho},I},\end{aligned}$$

that is,

$$(2.2.52)\qquad h_{18}(x) = \|\min\{\widetilde{\rho}(\cdot),\widetilde{\rho}(x)\}\|_{\infty,g/\widetilde{\rho},I} \text{ for all } x \in I.$$

Thus, the relation $h_{18} \in Q_{\widetilde{\rho}}(I)$ follows from Example 2.2.1 (ii).

(ii) Let $\rho \in Ads(I)$ and $w \in \mathcal{M}^+(I)$. Put

$$(2.2.53)\qquad \varphi(x) := \|\min\{\rho(\cdot),\rho(x)\}\|_{\infty,w,I},\ x \in I.$$

Then, by (2.2.51) and (2.2.43) (with $g = w$),

$$(2.2.54)\qquad \varphi(x) = \|\rho(t)\ \|w\|_{\infty,(t,b)}\|_{\infty,(a,x)},\ x \in I.$$

Moreover, by (2.2.52) and (2.2.50) (with $\widetilde{\rho} = \rho$ and $g = w\rho$),

$$(2.2.55)\qquad \varphi(x) = \rho(x)\left\|\frac{1}{\rho(t)}\|w\rho\|_{\infty,(a,t)}\right\|_{\infty,(x,b)},\ x \in I.$$

We make use of identity (2.2.55) to prove the next result which will be indispensable in Chapter 5.

Lemma 2.2.14 *Let $I = (a,b) \subseteq \mathbb{R}$, $\rho \in Ads(I)$, $f \in Q_\rho(I)$ and let $w \in \mathcal{W}(I)$. Define φ by (2.2.53) and assume that $\varphi \not\equiv +\infty$. Then*

$$(2.2.56)\qquad \|f\|_{\infty,w,I} = \|f\|_{\infty,\varphi/\rho,I}.$$

Proof. Since

$$\begin{aligned}\varphi(x) &= \max\{\|\rho w\|_{\infty,(a,x)},\ \rho(x)\|w\|_{\infty,[x,b)}\}\\ &\geq \rho(x)\|w\|_{\infty,[x,b)} \geq \rho(x)w(x) \quad \text{for a.e. } x \in I,\end{aligned}$$

we have

$$\frac{\varphi}{\rho} \geq w \quad \text{a.e. on } I, \tag{2.2.57}$$

which implies that LHS(2.2.56) $\leq$ RHS(2.2.56).

To prove the reverse inequality, we employ (2.2.55). Hence,

$$\left(\frac{\varphi}{\rho}\right)(x) = \left\|\frac{1}{\rho(t)}\|w\rho\|_{\infty,(a,t)}\right\|_{\infty,(x,b)} \quad \text{for all } x \in I. \tag{2.2.58}$$

On substituting (2.2.58) into the right-hand side of (2.2.56) and using the fact that $f \in \mathcal{M}^+(I;\uparrow)$, we obtain

$$\begin{aligned}\text{RHS(2.2.56)} &= \operatorname*{ess\,sup}_{x\in(a,b)} f(x) \operatorname*{ess\,sup}_{t\in(x,b)} \frac{1}{\rho(t)}\|w\rho\|_{\infty,(a,t)}\\ &\leq \operatorname*{ess\,sup}_{x\in(a,b)} \operatorname*{ess\,sup}_{t\in(x,b)} \frac{f(t)}{\rho(t)}\|w\rho\|_{\infty,(a,t)}.\end{aligned} \tag{2.2.59}$$

If we interchange the first two essential suprema on the right-hand side of (2.2.59), we arrive at

$$\begin{aligned}\text{RHS(2.2.59)} &= \operatorname*{ess\,sup}_{t\in(a,b)} \operatorname*{ess\,sup}_{x\in(a,t)} \frac{f(t)}{\rho(t)}\|w\rho\|_{\infty,(a,t)}\\ &= \left\|\frac{f(t)}{\rho(t)}\|w\rho\|_{\infty,(a,t)}\right\|_{\infty,I}.\end{aligned} \tag{2.2.60}$$

Finally, on interchanging the essential suprema in the last expression and applying the fact that $\frac{f}{\rho} \in \mathcal{M}^+(I;\downarrow) \cap C(I)$, we get

$$\text{RHS(2.2.60)} = \left\|w(s)\rho(s)\left\|\frac{f}{\rho}\right\|_{\infty,(s,b)}\right\|_{\infty,I} = \|w(s)f(s)\|_{\infty,I} = \text{LHS(2.2.56)}. \tag{2.2.61}$$

The desired inequality RHS(2.2.56) $\leq$ LHS(2.2.56) follows from (2.2.59)–(2.2.61). □

The following assertion is a modification of Lemmas 2.2.3 and 2.2.9.

Lemma 2.2.15 *Let $I = (a,b) \subseteq \mathbb{R}$, $G \in \mathcal{M}^+(I;\downarrow)$, $\rho \in Ads(I)$ and $\rho(a+) = 0$. Put*

$$F(x) := \frac{\|G\rho\|_{\infty,(a,x)}}{\rho(x)}, \quad x \in I. \tag{2.2.62}$$

Then

$$F \in \mathcal{M}^+(I;\downarrow) \tag{2.2.63}$$

and

$$F\rho \in Q_\rho(I). \tag{2.2.64}$$

Proof. If we put $H := G \circ \rho^{-1}$, then

$$H \in \mathcal{M}^+(J;\downarrow), \quad \text{where} \quad J = \rho(I) = (\rho(a+), \rho(b-)). \tag{2.2.65}$$

Moreover,

$$F(z) = \frac{\|H(\rho(t))\,\rho(t)\|_{\infty,(a,z)}}{\rho(z)} \quad \text{for all} \quad z \in I.$$

Thus, using the change of variables $s = \rho(t)$ and the fact that $\rho(a+) = 0$, we arrive at

$$F(z) = \frac{\|H(s)\ s\|_{\infty,(0,\rho(z))}}{\rho(z)} \quad \text{for all} \quad z \in I. \tag{2.2.66}$$

Now, let $x, y \in I$ and $x < y$. Then (2.2.66) and the change of variables $s = t\rho(x)/\rho(y)$ imply that

$$F(x) = \frac{\|H(s)\ s\|_{\infty,(0,\rho(x))}}{\rho(x)} = \frac{\|H(t\frac{\rho(x)}{\rho(y)})\ t\|_{\infty,(0,\rho(y))}}{\rho(y)}. \tag{2.2.67}$$

Since $t\rho(x)/\rho(y) \le t$ for all $t \ge 0$, we obtain from (2.2.65) and (2.2.66) that

$$\text{RHS(2.2.67)} \ge \frac{\|H(t)\ t\|_{\infty,(0,\rho(y))}}{\rho(y)} = F(y).$$

Together with (2.2.67), this yields (2.2.63).

Since, by (2.2.62), $F(x)\rho(x) = \|G\rho\|_{\infty,(a,x)}$, $x \in I$, it is clear that $F\rho \in \mathcal{M}^+(I;\uparrow)$. This relation and (2.2.63) yield (2.2.64). □

The next assertion is a counterpart of Lemma 2.2.15 (cf. also Lemma 2.2.11).

Lemma 2.2.16 *Let $I = (a,b) \subseteq \mathbb{R}$, $G \in \mathcal{M}^+(I;\uparrow)$, $\widetilde{\rho} \in Ads(I)$ and $\widetilde{\rho}(b-) = +\infty$. Put*

$$F(x) := \widetilde{\rho}(x)\|G/\widetilde{\rho}\|_{\infty,(x,b)}, \quad x \in I. \tag{2.2.68}$$

Then

$$F \in \mathcal{M}^+(I;\uparrow) \tag{2.2.69}$$

and

$$F \in Q_{\widetilde{\rho}}(I). \tag{2.2.70}$$

Proof. Putting $\Phi(x) = -1/\widetilde{\rho}(x)$, $x \in I$, we see that the function $H := G \circ \Phi^{-1}$ satisfies

$$H \in \mathcal{M}^+(J;\uparrow), \text{ where } J = (\Phi(a+), \Phi(b-)). \tag{2.2.71}$$

Moreover,

$$F(z) = \frac{\|H(\Phi(t))(-\Phi(t))\|_{\infty,(z,b)}}{-\Phi(z)} \quad \text{for all } z \in I.$$

Thus, using the change of variables $s = \Phi(t)$ and the fact that $\Phi(b-) = 0$, we arrive at

$$F(z) = \frac{\|H(s)(-s)\|_{\infty,(\Phi(z),0)}}{-\Phi(z)} \quad \text{for all } z \in I. \tag{2.2.72}$$

Now, let $x, y \in I$ and $x < y$. Then (2.2.72) and the change of variables $s = t\Phi(x)/\Phi(y)$ imply that

$$F(x) = \frac{\|H(s)(-s)\|_{\infty,(\Phi(x),0)}}{-\Phi(x)} = \frac{\|H(t\frac{\Phi(x)}{\Phi(y)})(-t)\|_{\infty,(\Phi(y),0)}}{-\Phi(y)}. \tag{2.2.73}$$

Since $t\Phi(x)/\Phi(y) \le t$ for all $t \le 0$, we obtain from (2.2.71) and (2.2.72) that

$$\text{RHS}(2.2.73) \le \frac{\|H(t)(-t)\|_{\infty,(\Phi(y),0)}}{-\Phi(y)} = F(y).$$

Together with (2.2.73), this yields (2.2.69).

Since, by (2.2.68) $F(x)/\widetilde{\rho}(x) = \|G(t)/\widetilde{\rho}(t)\|_{\infty,(x,b)}$, $x \in I$, it is clear that $F/\widetilde{\rho} \in \mathcal{M}^+(I;\downarrow)$. This relation and (2.2.69) yield (2.2.70). □

The following two results generalize those of Lemma 2.2.3 and Lemma 2.2.6.

Lemma 2.2.17 *Let μ be a non-negative Borel measure on the interval $I = (a,b) \subseteq \mathbb{R}$ such that the function $\rho(x) := \mu((a,x])$, $x \in I$, is continuous and strictly increasing on I. Assume that $G \in \mathcal{M}^+(I;\downarrow)$. Put*

$$F(x) := \frac{\int_{(a,x]} G\,d\mu}{\mu((a,x])}, \quad x \in I. \tag{2.2.74}$$

Then

$$F \in \mathcal{M}^+(I;\downarrow) \tag{2.2.75}$$

and

$$F\rho \in Q_\rho(I). \tag{2.2.76}$$

Proof. If $a < x \le y < b$, then, on using the monotonicity of G, we obtain

$$F(x) = \frac{\mu((a,x]) + \mu((x,y])}{\mu((a,y])} \frac{\int_{(a,x]} G\,d\mu}{\mu((a,x])} \tag{2.2.77}$$
$$= \frac{\int_{(a,x]} G\,d\mu}{\mu((a,y])} + \frac{\mu((x,y]) \int_{(a,x]} G\,d\mu}{\mu((a,y])\,\mu((a,x])}$$
$$\ge \frac{\int_{(a,x]} G\,d\mu}{\mu((a,y])} + \frac{\mu((x,y])}{\mu((a,y])} G(x)$$
$$\ge \frac{\int_{(a,x]} G\,d\mu}{\mu((a,y])} + \frac{\int_{(x,y]} G\,d\mu}{\mu((a,y])}$$
$$= F(y).$$

Consequently, (2.2.75) holds. Moreover,

$$F(x)\rho(x) = \int_{(a,x]} G\,d\mu, \quad x \in I,$$

and hence $F\rho \in \mathcal{M}^+(I;\uparrow)$. Since $\rho \in C(I) \cap \mathcal{M}_s^+(I;\uparrow)$, we have $\rho \in Ads(I)$. Summarizing these results, we see that (2.2.76) is satisfied. □

Lemma 2.2.18 *Let μ be a non-negative Borel measure on the interval $I = (a,b) \subseteq \mathbb{R}$ such that the function $x \mapsto \mu([x,b))$, $x \in I$, is continuous and strictly decreasing on I. Assume that $G \in \mathcal{M}^+(I;\uparrow)$. Put*

$$F(x) := \frac{\int_{[x,b)} G\,d\mu}{\mu([x,b))}, \quad x \in I. \tag{2.2.78}$$

Then

$$F \in \mathcal{M}^+(I;\uparrow) \tag{2.2.79}$$

and

$$F \in Q_{\widetilde{\rho}}(I), \tag{2.2.80}$$

where

$$\widetilde{\rho}(x) := \frac{1}{\mu([x,b))}, \quad x \in I.$$

Proof. Since the proof is analogous to that of Lemma 2.2.17, it is left to the reader. □

Definition 2.2.19 Let $I = (a,b) \subseteq \mathbb{R}$ and let $T : \mathcal{M}^+(I) \to \mathcal{M}^+(I)$ be an operator with a domain $\mathcal{D}(T)$. We say that T is *ρ-quasiconcave operator* if there exists $\rho \in Ads(I)$ such that $Tf \in Q_\rho(I)$ for all $f \in \mathcal{D}(T)$.

Examples 2.2.20 Let $I = (a,b) \subseteq \mathbb{R}$ and $u \in \mathcal{W}(I)$ be such that the function ρ given by $\rho(x) = \int_a^x u(t)\,dt$, $x \in I$, belongs to $Ads(I)$. Then the operator T defined on $\mathcal{M}^+(I)$ by

$$(Tf)(x) = \int_a^x u(t)\Big(\int_t^b f(s)\,ds\Big)\,dt, \quad x \in I,$$

is ρ-quasiconcave, since, by Corollary 2.2.4,

$$Tf \in Q_\rho(I) \text{ for every } f \in \mathcal{M}^+(I).$$

On the other hand, the operator T defined on $\mathcal{D}(T) := \mathcal{M}^+(I;\uparrow)$ by $Tf = f$ is not ρ-quasiconcave. Indeed, assuming the opposite, we have $Tf \in Q_\rho(I)$ for some $\rho \in Ads(I)$ and all $f \in \mathcal{M}^+(I;\uparrow)$. But taking $f = \rho^2$, we arrive at a contradiction since $\frac{T\rho^2}{\rho} = \rho \notin \mathcal{M}^+(I;\downarrow)$.

Another example of an operator T which is not ρ-quasiconcave is the following: Let $T : \mathcal{M}^+(I) \to M^+(I)$, with $\mathcal{D}(I) = \mathcal{M}^+(I)$, be given by $(Tf)(x) = \int_a^x f(t)\,dt$, $x \in I$.

To verify that T is not ρ-quasiconcave, take any $\rho \in Ads(I)$, $x_1, x_2 \in I$ with $x_1 < x_2$, and $f = \chi_{(x_1,b)}$. Then

$$\Big(\frac{Tf}{\rho}\Big)(x_1) = 0 < \frac{x_2 - x_1}{\rho(x_2)} = \Big(\frac{Tf}{\rho}\Big)(x_2).$$

Consequently, $\frac{Tf}{\rho} \notin \mathcal{M}^+(I;\downarrow)$ and the result follows.

2.3 The ρ-fundamental function of $L^p(w, I, \mu)$

The ρ-quasiconcave function h_7 in (2.2.4) is of particular importance, which merits the following special name.

Definition 2.3.1 Let μ be a non-negative Borel measure on the interval $I = (a,b) \subseteq \mathbb{R}$, $\rho \in Ads(I)$, $w \in \mathcal{W}(I)$, and $p \in (0, +\infty]$.

(i) The function $\varphi \in Q_\rho(I)$ associated to the space $L^p(w, I, \mu)$ and defined by

$$\varphi(x) := \|\min\{\rho(\cdot), \rho(x)\}\|_{p,w,I,\mu}, \quad x \in I, \tag{2.3.1}$$

is called the *ρ-fundamental function* of the space $L^p(w, I, \mu)$.

(ii) The space $L^p(w, I, \mu)$ is called *ρ-non-trivial* if its ρ-fundamental function φ satisfies $0 \not\equiv \varphi \not\equiv +\infty$.

(iii) Assume, moreover, that

$$\lim_{x \to a+} \rho(x) = 0 \quad \text{and} \quad \lim_{x \to b-} \rho(x) = +\infty. \tag{2.3.2}$$

Then a ρ-non-trivial space $L^p(w, I, \mu)$, $p \in (0, +\infty)$, is said to be *ρ-non-degenerate* if, for some $c \in I$,

$$\|\chi_{(a,c)}\|_{p,w,I,\mu} = +\infty \quad \text{and} \quad \|\rho\chi_{[c,b)}\|_{p,w,I,\mu} = +\infty. \tag{2.3.3}$$

In Chapter 5 we shall need the following result.

Lemma 2.3.2 *Let μ be a non-negative Borel measure on the interval $I = (a,b) \subseteq \mathbb{R}$, $\rho \in Ads(I)$, $w \in \mathcal{W}(I)$, $p \in (0, +\infty]$, and let φ be the ρ-fundamental function of the space $L^p(w, I, \mu)$. Then*

$$\lim_{x \to a+} \left(\frac{\rho}{\varphi}\right)(x) = \frac{1}{\|w\|_{p,I,\mu}} \tag{2.3.4}$$

and

$$\lim_{x \to b-} \varphi(x) = \|\rho w\|_{p,I,\mu}. \tag{2.3.5}$$

Proof. Assume first that $p \in (0, +\infty)$.
Since

$$L_1 := \lim_{x \to a+} \left(\frac{\rho}{\varphi}\right)(x) = \lim_{x \to a+} \frac{1}{\left(\int_{(a,x]} \left(\frac{\rho(t)}{\rho(x)}\right)^p w^p(t)\, d\mu(t) + \int_{(x,b)} w^p(t)\, d\mu(t)\right)^{1/p}},$$

we obtain

$$L_1 \leq \lim_{x \to a+} \frac{1}{\left(\int_{(x,b)} w^p(t)\, d\mu(t)\right)^{1/p}} = \frac{1}{\|w\|_{p,I,\mu}}$$

and

$$L_1 \geq \lim_{x \to a+} \frac{1}{\left(\int_{(a,x]} w^p(t)\, d\mu(t) + \int_{(x,b)} w^p(t)\, d\mu(t)\right)^{1/p}} = \frac{1}{\|w\|_{p,I,\mu}}.$$

Consequently, (2.3.4) holds.
Similarly, as

$$L_2 := \lim_{x \to b-} \varphi(x) = \lim_{x \to b-} \left(\int_{(a,x]} \rho^p(t)\, w^p(t)\, d\mu(t) + \int_{(x,b)} \rho^p(x)\, w^p(t)\, d\mu(t)\right)^{1/p},$$

we arrive at

$$L_2 \leq \lim_{x \to b-} \left(\int_{(a,x]} \rho^p(t)\, w^p(t)\, d\mu(t) + \int_{(x,b)} \rho^p(t)\, w^p(t)\, d\mu(t)\right)^{1/p} = \|\rho w\|_{p,I,\mu}$$

and

$$L_2 \geq \lim_{x \to b-} \left(\int_{(a,x]} \rho^p(t)\, w^p(t)\, d\mu(t)\right)^{1/p} = \|\rho w\|_{p,I,\mu}.$$

Therefore, (2.3.5) is satisfied.
If $p = +\infty$, then the proof can be done analogously and thus we leave it to the reader. □

A counterpart of the previous lemma is the following assertion.

Lemma 2.3.3 *Let μ be a non-negative Borel measure on the interval $I = (a,b) \subseteq \mathbb{R}$, $\rho \in Ads(I)$, $w \in \mathcal{W}(I)$, $p \in (0,+\infty)$, and let φ be the ρ-fundamental function of the space $L^p(w,I,\mu)$. If $\varphi \not\equiv +\infty$, then* [3]

$$\lim_{x\to a+} \varphi(x) = \rho(a+)\,\|w\|_{p,I,\mu} \tag{2.3.6}$$

and

$$\lim_{x\to b-} \left(\frac{\varphi}{\rho}\right)(x) = \frac{\|\rho w\|_{p,I,\mu}}{\rho(b-)}. \tag{2.3.7}$$

Proof. To verify (2.3.6), consider a non-increasing sequence $\{x_n\}_{n\in\mathbb{N}}$ of points $x_n \in I$, $n \in \mathbb{N}$, satisfying $x_n \to a$. Then

$$\varphi(x_n) = \|f_n\|_{p,w,I,\mu},$$

where

$$f_n(t) := \min\{\rho(t), \rho(x_n)\}, \quad t \in I, \quad n \in \mathbb{N}.$$

Obviously, $f_{n+1} \le f_n$, $f_n \in \mathcal{M}^+(I,\uparrow)$, $n \in \mathbb{N}$, and, for any fix $t \in I$,

$$\lim_{n\to+\infty} f_n(t) = \lim_{n\to+\infty} \rho(x_n) = \rho(a+).$$

Since $\varphi \not\equiv +\infty$, we have $\varphi(x) < +\infty$ for all $x \in I$. In particular, $\varphi(x_1) < +\infty$, which means that $(f_1 w)^p \in L^1(I,\mu)$. Thus, Lebesgue's dominated convergence theorem implies that

$$\lim_{n\to+\infty} \varphi^p(x_n) = \lim_{n\to+\infty} \int_I [f_n(t)w(t)]^p \, d\mu = \int_I [\rho(a+)w(t)]^p \, d\mu.$$

Consequently,

$$\lim_{n\to+\infty} \varphi(x_n) = \|\rho(a+)w(\cdot)\|_{p,I,\mu} = \rho(a+)\,\|w\|_{p,I,\mu}$$

(the last equality is always true due to the convention that $0.(+\infty) = 0$). Together with the fact that $\varphi \in \mathcal{M}^+(I;\uparrow)$, this shows that (2.3.6) holds.

To verify (2.3.7), consider a non-decreasing sequence $\{y_n\}_{n\in\mathbb{N}}$ of points $y_n \in I, n \in \mathbb{N}$, satisfying $y_n \to b$. Then

$$\left(\frac{\varphi}{\rho}\right)(y_n) = \|g_n\|_{p,w,I,\mu},$$

where

$$g_n(t) := \min\left\{\frac{\rho(t)}{\rho(y_n)}, 1\right\}, \quad t \in I, \quad n \in \mathbb{N}.$$

[3] Note that in (2.3.6) we use the convention that $0.(+\infty) = 0$ - cf. Convention 1.1.1 (i). Similarly, in (2.3.7) we use the convention that $(+\infty)/(+\infty) = 0$.

Obviously, $g_{n+1} \le g_n$, $g_n \in \mathcal{M}^+(I; \uparrow)$, $n \in \mathbb{N}$, and, for any fix $t \in I$,

$$\lim_{n\to+\infty} g_n(t) = \lim_{n\to+\infty} \frac{\rho(t)}{\rho(y_n)} = \frac{\rho(t)}{\rho(b-)}.$$

Since $\varphi \not\equiv +\infty$, we have $(\frac{\varphi}{\rho})(y_1) < +\infty$, which means that $(g_1 w)^p \in L^1(I, \mu)$. Thus, Lebesgue's dominated convergence theorem implies that

$$\lim_{n\to+\infty} \left(\frac{\varphi}{\rho}\right)^p (y_n) = \lim_{n\to+\infty} \int_I [g_n(t)w(t)]^p \, d\mu = \int_I \left[\frac{\rho(t)}{\rho(b-)} w(t)\right]^p d\mu.$$

Consequently,

$$\lim_{n\to+\infty} \left(\frac{\varphi}{\rho}\right)(y_n) = \left\|\frac{\rho(\cdot)w(\cdot)}{\rho(b-)}\right\|_{p,I,\mu} = \frac{\|\rho w\|_{p,I,\mu}}{\rho(b-)}$$

(the last equality is always true due to the convention that $(+\infty)/(+\infty) = 0$). Together with the fact that $\frac{\varphi}{\rho} \in \mathcal{M}^+(I; \uparrow)$, this shows that (2.3.7) is satisfied. □

Remark 2.3.4 If $p = +\infty$ and φ is the *ρ-fundamental function* of the space $L^p(w, I, \mu)$, then it follows from Lemma 2.3.2 that

$$\lim_{x\to a+} \varphi(x) = \lim_{x\to a+} \frac{\rho(x)}{\left(\frac{\rho}{\varphi}\right)(x)} = \frac{\lim\limits_{x\to a+} \rho(x)}{\lim\limits_{x\to a+} \left(\frac{\rho}{\varphi}\right)(x)} = \rho(a+) \, \|w\|_{p,I,\mu}$$

provided that $\rho(a+) \neq 0$ or $\|w\|_{p,I,\mu}) \neq +\infty$, and that

$$\lim_{x\to b-} \left(\frac{\varphi}{\rho}\right)(x) = \frac{\lim\limits_{x\to b-} \varphi(x)}{\lim\limits_{x\to b-} \rho(x)} = \frac{\|\rho w\|_{p,I,\mu}}{\rho(b-)}$$

provided that $\rho(b-) \neq +\infty$ or $\|\rho w\|_{p,I,\mu} \neq +\infty$.

Lemma 2.3.5 *Let μ be a non-negative Borel measure on the interval $I = (a, b) \subseteq \mathbb{R}$, $\rho \in Ads(I)$, $w \in \mathcal{W}(I)$, $p \in (0, +\infty)$, and let the space $L^p(w, I, \mu)$ be ρ-non-degenerate. Then the ρ-fundamental function φ of the space $L^p(w, I, \mu)$ belongs to $Q^0_\rho(I)$, that is, $\varphi \in Q_\rho(I)$ and*

$$\text{(2.3.8)} \qquad \lim_{x\to a+} \varphi(x) = 0, \qquad \lim_{x\to b-} \varphi(x) = +\infty,$$

$$\text{(2.3.9)} \qquad \lim_{x\to a+} \left(\frac{\rho}{\varphi}\right)(x) = 0, \qquad \lim_{x\to b-} \left(\frac{\rho}{\varphi}\right)(x) = +\infty.$$

Proof. Since the space $L^p(w, I, \mu)$ is ρ-non-degenerate, there is $c \in I$ such that (2.3.3) holds. Therefore,

$$\|w\|_{p,I,\mu} = +\infty \qquad \text{and} \qquad \|\rho w\|_{p,I,\mu} = +\infty.$$

Together with Lemma 2.3.2, this implies that the first condition in (2.3.9) and the second one in (2.3.8) are satisfied. Moreover, since the space $L^p(w, I, \mu)$ is ρ-non-degenerate, condition (2.3.2) holds and $\varphi \not\equiv +\infty$. These facts and Lemma 2.3.3 then guarantee that the remaining conditions in (2.3.8) and (2.3.9) are satisfied as well. □

Remark 2.3.6 We have assumed (2.3.2) in the definition of the ρ-non-degenerate space $L^p(w, I, \mu)$, $p \in (0, +\infty)$, and for such a space we have proved (2.3.8) and (2.3.9), that is, we have shown that $\varphi \in Q^0_\rho(I)$.

Suppose on the other hand that φ is any element of the class $Q^0_\rho(I)$. Then (by the definition of $Q^0_\rho(I)$), (2.3.8) and (2.3.9) are satisfied. But the first conditions in (2.3.8) and (2.3.9) imply that necessarily $\lim_{x\to a+} \rho(x) = 0$, which is the first condition in (2.3.2). Analogously, the second conditions in (2.3.8) and (2.3.9) yield the second condition in (2.3.2). This shows that assumption (2.3.2) in the definition of ρ-non-degenerate space is quite natural. (Moreover, it follows from our discussion that it is natural to speak about the subclass $Q^0_\rho(I)$ as the class of all non-degenerate ρ-quasiconcave functions on I.)

2.4 Representation of $Q_\rho(I)$-functions

By Example 2.2.1 (ii), if $\rho \in Ads(I)$ and μ is a non-negative Borel measure on the interval $I = (a, b) \subseteq \mathbb{R}$, then the function

$$h(t) = \int_I \min\{\rho(s), \rho(t)\}\, d\mu(s), \ t \in I,$$

belongs to $Q_\rho(I)$. Now, we are going to show that any function $h \in Q_\rho(I)$ can be represented in a similar form.

Theorem 2.4.1 *Let $I = (a, b) \subseteq \mathbb{R}$, $\rho \in Ads(I)$ and let $h \in Q_\rho(I)$. Then there is a non-negative Borel measure μ on I such that, for all $t \in I$,*

$$h(t) \le \alpha + \beta\rho(t) + \int_I \min\{\rho(s), \rho(t)\}\, d\mu(s) \le 4h(t), \tag{2.4.1}$$

where

$$\alpha = \lim_{t\to a+} h(t) \quad and \quad \beta = \lim_{t\to b-} \frac{h(t)}{\rho(t)}. \tag{2.4.2}$$

In particular,

$$h(t) \approx \alpha + \beta\rho(t) + \int_I \min\{\rho(s), \rho(t)\}\, d\mu(s) \quad for\ all\ t \in I.$$

To prove Theorem 2.4.1, we shall need the following lemma.

Lemma 2.4.2 *Let $0 < A < B \le +\infty$ and $J = (A, B)$. Suppose that $g \in Q(J)$ and that G is the least concave majorant of g. Then*

$$g \le G \le 2g \text{ on } J, \tag{2.4.3}$$

$$G \in \mathcal{M}^+(J; \uparrow), \tag{2.4.4}$$

$$G(A+) = g(A+), \tag{2.4.5}$$

$$\lim_{t \to B-} \frac{G(t)}{t} = \lim_{t \to B-} \frac{g(t)}{t}. \tag{2.4.6}$$

Proof of Lemma 2.4.2. The estimate (2.4.3) holds by Remark 2.2.2 (ii) (cf. (2.2.12)).

Assume that (2.4.4) is not satisfied. Then there are $x_1, x_2 \in (A, B)$, $x_1 < x_2$, such that $G(x_1) > G(x_2)$. The concavity of G on (A, B) implies that

$$G(x) < G(x_2) \quad \text{if} \quad x \in (x_2, B) \tag{2.4.7}$$

(otherwise the point $[x_2, G(x_2)]$ lies underneath the line segment joining $[x_1, G(x_1)]$ and $[x, G(x)]$). Together with the fact that $g \in \mathcal{M}^+(J; \uparrow)$, this implies that the constant function $f(x) = G(x_2)$, $x \in J$, dominates g on J. Therefore, the concave function $\widetilde{G} := \min\{G, f\}$ dominates g on J as well. However, this is a contradiction since $\widetilde{G}(x_1) < G(x_1)$. Consequently, (2.4.4) holds.

To prove (2.4.5), assume on the contrary that $G(A+) > g(A+)$ (as G dominates g on J, it is clear that $G(A+) \ge g(A+)$). Then, since $G, g \in C(J)$, there is $D \in (A, B)$ such that

$$g(x) < \frac{G(A+) + g(A+)}{2} \quad \text{for all } x \in (A, D].$$

Let $\widetilde{G}$ be the function on (A, B) whose graph is the union of $G|[D, B)$ (the restriction of G to $[D, B)$) and the open line segment with endpoints $[A, \frac{G(A+)+g(A+)}{2}]$, $[D, G(D)]$. Then $\widetilde{G}$ is concave on (A, B), $\widetilde{G}$ dominates g on (A, B) and $\widetilde{G}(x) < G(x)$ for all $x \in (A, D)$. This is a contradiction since G is the least concave majorant of g on (A, B). Thus, (2.4.5) is proved.

To verify (2.4.6), it is enough to prove that

$$\limsup_{t \to B-} \frac{G(t)}{t} \le \lim_{t \to B-} \frac{g(t)}{t}. \tag{2.4.8}$$

Indeed, using (2.4.8) and the facts that $g \le G$ on J and $g \in Q(J)$, we obtain

$$\lim_{t \to B-} \frac{g(t)}{t} \ge \limsup_{t \to B-} \frac{G(t)}{t} \ge \liminf_{t \to B-} \frac{G(t)}{t} \ge \liminf_{t \to B-} \frac{g(t)}{t} = \lim_{t \to B-} \frac{g(t)}{t},$$

and (2.4.6) follows.

To prove (2.4.8), we distinguish two cases:

First, let $B < +\infty$. We claim that

$$G(B-) = g(B-). \tag{2.4.9}$$

Suppose on the contrary that

$$G(B-) > g(B-) \tag{2.4.10}$$

(since G dominates g on (A, B), it is clear that $G(B-) \geq g(B-)$). As $g \in \mathcal{M}^+(J; \uparrow)$, we have $g(t) \leq g(B-)$ on (A, B). Consequently, the concave function $\widetilde{G}(t) := \min\{G(t), g(B-)\}$, $t \in (A, B)$, dominates g on (A, B) and satisfies (cf. (2.4.10)) $\widetilde{G}(t) < G(t)$ for all $t \in (A, B)$ which are close enough to B. But this contradicts the fact that G is the least concave majorant of g on (A, B). Therefore, (2.4.9) holds. Now, using the inequality $B < +\infty$, (2.4.4), (2.4.9) and the relation $g \in Q(J)$, we obtain

$$\limsup_{t\to B-} \frac{G(t)}{t} = \frac{G(B-)}{B} = \frac{g(B-)}{B} = \lim_{t\to B-} \frac{g(t)}{t}, \tag{2.4.11}$$

which means that (2.4.8) holds.

Assume now that $B = +\infty$. Let $x \in J$. Since $g \in Q(J)$, we have (cf. (2.1.7))

$$g(t) \leq \Big(1 + \frac{t}{x}\Big) g(x) \quad \text{for all } t \in J.$$

The function $t \mapsto (1 + \frac{t}{x}) g(x)$ is concave and dominates g on J. Hence, $G(t) \leq (1 + \frac{t}{x}) g(x)$ for all $t \in J$, which implies that

$$\limsup_{t\to+\infty} \frac{G(t)}{t} \leq \limsup_{t\to+\infty} \Big(\frac{1}{t} + \frac{1}{x}\Big) g(x) = \frac{g(x)}{x}.$$

Consequently, (2.4.8) holds. □

Proof of Theorem 2.4.1. Let $A = \rho(a+) = \lim_{t\to a+} \rho(t)$ and $B = \rho(b-) = \lim_{t\to b-} \rho(t)$. Then $\rho(I) = \{\rho(t); t \in I\} = (A, B)$. Without loss of generality we can assume that

$$0 \not\equiv h \not\equiv +\infty. \tag{2.4.12}$$

The function g defined by

$$g(t) = h(\rho^{-1}(t)), \quad t \in (A, B), \tag{2.4.13}$$

belongs to $Q(\rho(I))$ (cf. (2.1.11)). If G is the least concave majorant of g and $J = (A, B)$, then Lemma 2.4.2 shows that (2.4.3)–(2.4.6) are satisfied.

Since G is concave on (A, B) and (2.4.4) holds, Remark 2.2.2 (iii) implies that, for all $x, y \in (A, B)$ with $x < y$,

$$G(y) = G(x) + \int_x^y \theta(\sigma)\, d\sigma, \tag{2.4.14}$$

where θ is a non-negative, finite, non-increasing and right continuous function on (A, B). Using the integration by parts for Riemann-Stieltjes integrals (cf., e.g., [R, Theorem 6.30]), we obtain from (2.4.14) that

$$G(y) = G(x) + y\theta(y) - x\theta(x) + \int_x^y \sigma\, d(-\theta(\sigma)).^4 \tag{2.4.15}$$

Let $z \in (A, B)$, $z > y$. Then, by (2.4.15),

$$\begin{aligned} G(y) &= G(x) + y[\theta(y) - \theta(z)] + y\theta(z) - x\theta(x) + \int_x^y \sigma\, d(-\theta(\sigma)) \\ &= G(x) + y \int_y^z d(-\theta(\sigma)) + y\theta(z) - x\theta(x) + \int_x^y \sigma\, d(-\theta(\sigma)) \\ &= G(x) - x\theta(x) + y\theta(z) + \int_x^z \min\{y, \sigma\}\, d(-\theta(\sigma)). \end{aligned}$$

Since any point $y \in (A, B)$ is of the form $y = \rho(t)$, $t \in I$, we have

$$G(\rho(t)) = G(x) - x\theta(x) + \rho(t)\theta(z) + \int_x^z \min\{\rho(t), \sigma\}\, d(-\theta(\sigma))$$

if $x < \rho(t) < z$. Using the change of variables $\sigma = \rho(s)$ (cf., e.g., [Fe, 2.5.18 (2)]), we arrive at

$$G(\rho(t)) = G(x) - x\theta(x) + \rho(t)\theta(z) + \int_{\rho^{-1}(x)}^{\rho^{-1}(z)} \min\{\rho(t), \rho(s)\}\, d(-\theta(\rho(s))).$$

Define now a non-negative Borel measure μ on I as the unique extension of the function μ given by

$$\mu((c, d]) = \theta(\rho(c)) - \theta(\rho(d)) \quad \text{if} \quad c, d \in I,\ c < d.^5 \tag{2.4.16}$$

Then (cf. Remark 1.2.1)

$$\int_{\rho^{-1}(x)}^{\rho^{-1}(z)} \min\{\rho(t), \rho(s)\}\, d(-\theta(\rho(s))) = \int_{(\rho^{-1}(x), \rho^{-1}(z)]} \min\{(\rho(t), \rho(s)\}\, d\mu(s).$$

Therefore,

$$G(\rho(t)) = G(x) - x\theta(x) + \rho(t)\theta(z) + \int_{(\rho^{-1}(x), \rho^{-1}(z)]} \min\{\rho(t), \rho(s)\}\, d\mu(s)$$

---

[4]This is a Riemann-Stieltjes integral.

[5]Note that this definition is consistent with the right continuity of θ on J; recall that ρ is a continuous function on I.

for every $x, z \in (A, B)$ and all $t \in I$ such that $x < \rho(t) < z$. Consequently, for all $t \in I$,

$$G(\rho(t)) = G(A+) - \lim_{x \to A+} x\theta(x) + \rho(t) \lim_{z \to B-} \theta(z) + \int_{(a,b)} \min\{\rho(t), \rho(s)\}\, d\mu(s). \tag{2.4.17}$$

Next we prove that

$$\theta(y) \le \frac{G(y)}{y} \quad \text{for all } y \in (A, B). \tag{2.4.18}$$

Indeed, by (2.4.14), we have

$$G(y) - G(A+) = \int_A^y \theta(\sigma)\, d\sigma \ge \theta(y)(y - A), \quad y \in (A, B),$$

which implies that

$$\theta(y) \le \frac{G(y) - G(A+)}{y - A} \quad \text{for all } y \in (A, B). \tag{2.4.19}$$

If $A = 0$, then (2.4.18) follows from (2.4.19) since $G(A+) \ge 0$. Assume that $A > 0$. Then, since $g \in Q((A, B))$,

$$g(y) \le \frac{g(x)}{x} y \quad \text{for all } x, y \in (A, B),\ x < y.$$

Hence,

$$g(y) \le \frac{g(A+)}{A} y \quad \text{for all } y \in (A, B).$$

(Note that (2.4.12) yields $g(A+) < +\infty$, which, together with the condition $A > 0$, shows that $g(A+)/A < +\infty$.) This implies that

$$G(y) \le \frac{g(A+)}{A} y \quad \text{for all } y \in (A, B). \tag{2.4.20}$$

By (2.4.5) and (2.4.20),

$$\frac{G(y)}{y} A \le G(A+) \quad \text{for all } y \in (A, B).$$

Therefore,

$$\text{RHS}(2.4.19) \le \frac{G(y) - AG(y)/y}{y - A} = \frac{G(y)}{y} \quad \text{for all } y \in (A, B),$$

and (2.4.18) again follows.

By (2.4.18), (2.4.6), (2.4.13) and (2.4.2),

$$\lim_{z\to B-}\theta(z) \le \lim_{z\to B-}\frac{G(z)}{z} = \lim_{z\to B-}\frac{g(z)}{z} = \beta. \tag{2.4.21}$$

Together with the fact that $\theta \ge 0$, relations (2.4.21), (2.4.5), (2.4.17), (2.4.13) and (2.4.2) imply that, for all $t \in I$,

$$G(\rho(t)) \le \alpha + \beta\rho(t) + \int_I \min\{\rho(t), \rho(s)\}\, d\mu(s).$$

Moreover, using (2.4.3) and (2.4.13), we arrive at

$$h(t) \le \alpha + \beta\rho(t) + \int_I \min\{\rho(t), \rho(s)\}\, d\mu(s) \quad \text{if} \quad t \in I,$$

and the first inequality in (2.4.1) follows.

Now we prove the second inequality in (2.4.1). By (2.4.18)

$$G(A+) - \lim_{x\to A+} x\theta(x) \ge G(A+) - \lim_{x\to A+} G(x) = 0.$$

This estimate, the inequality $\theta \ge 0$ and (2.4.17) imply that, for all $t \in I$,

$$G(\rho(t)) \ge \int_I \min\{\rho(t), \rho(s)\}\, d\mu(s).$$

Therefore, using also (2.4.3) and (2.4.13), we obtain that, for all $t \in I$,

$$h(t) = g(\rho(t)) \ge \frac{1}{2}G(\rho(t)) \ge \frac{1}{2}\int_I \min\{\rho(t), \rho(s)\}\, d\mu(s).$$

Furthermore, since $h \in Q_\rho(I)$, we have

$$h(t) \ge h(a+) = \alpha \quad \text{and} \quad \frac{h(t)}{\rho(t)} \ge \beta \quad \text{for all } t \in I.$$

Consequently,

$$4h(t) \ge \alpha + \beta\rho(t) + \int_I \min\{\rho(t), \rho(s)\}\, d\mu(s) \quad \text{for all } t \in I,$$

which is the second inequality in (2.4.1). □

By Theorem 2.4.1, any $h \in Q_\rho(I)$ has a representation of the form

$$h(x) \approx \alpha + \beta\rho(x) + \int_I \min\{\rho(x), \rho(t)\}\, d\mu(t) \quad \text{for all} \quad x \in I \tag{2.4.22}$$

with some $\mu \in \mathcal{B}^+(I)$ and α, β from (2.4.2). On the other hand, Lemma 2.1.3 shows that, for any $p \in (0, +\infty)$,

$$h^p \in Q_{\rho^p}(I) \quad \text{and} \quad \left(\frac{\rho}{h}\right)^p \in Q_{\rho^p}(I).$$

In this connection a natural question arises: Are we able to determine representations of h^p and $(\rho/h)^p$ by means of (2.4.22)? The following two theorems solve this problem.

Theorem 2.4.3 *Let* $I = (a, b) \subseteq \mathbb{R}$, $\rho \in Ads(I)$, $h \in Q_\rho(I)$, $\alpha, \beta \in [0, +\infty]$, $\mu \in \mathcal{B}^+(I)$, $p \in (0, +\infty)$ *and let*

$$h(x) \approx \alpha + \beta\rho(x) + \int_I \min\{\rho(x), \rho(t)\}\, d\mu(t) \quad \text{for all} \quad x \in I. \tag{2.4.23}$$

Then, for all $x \in I$,

$$h^p(x) \approx \alpha^p + \beta^p\rho^p(x) + \int_I \min\{\rho^p(x), \rho^p(t)\}\Big(\frac{H}{\rho}\Big)^{p-1}(t)\, d\mu(t), \tag{2.4.24}$$

where

$$H(x) = \int_I \min\{\rho(x), \rho(t)\}\, d\mu(t), \quad x \in I. \tag{2.4.25}$$

Proof. Obviously, it is sufficient to prove that

$$H^p(x) \approx \int_a^b \min\{\rho^p(x), \rho^p(t)\}\Big(\frac{H}{\rho}\Big)^{p-1}(t)\, d\mu(t) \quad \text{for all} \quad x \in I. \tag{2.4.26}$$

Let $0 < \varepsilon \le \min\{1, p\}$. Then, since $H^{p-\varepsilon} \in \mathcal{M}^+(I;\uparrow)$ and $(H/\rho)^{p-\varepsilon} \in \mathcal{M}^+(I;\downarrow)$, we have for all $x \in I$,

$$\begin{aligned} \text{RHS(2.4.26)} &\le \int_{(a,x]} H^{p-1}(t)\rho(t)\, d\mu(t) \\ &\quad + \rho^p(x)\int_{[x,b)} \Big(\frac{H}{\rho}\Big)^{p-1}(t)\, d\mu(t) \\ &= \int_{(a,x]} H^{p-\varepsilon}(t)H^{\varepsilon-1}(t)\rho(t)\, d\mu(t) \\ &\quad + \rho^p(x)\int_{[x,b)} \Big(\frac{H}{\rho}\Big)^{p-\varepsilon}(t)\Big(\frac{H}{\rho}\Big)^{\varepsilon-1}(t)\, d\mu(t) \\ &\le H^{p-\varepsilon}(x)\int_{(a,x]} H^{\varepsilon-1}(t)\rho(t)\, d\mu(t) \\ &\quad + \Big(\frac{H}{\rho}\Big)^{p-\varepsilon}(x)\rho^p(x)\int_{[x,b)} \Big(\frac{H}{\rho}\Big)^{\varepsilon-1}(t)\, d\mu(t). \end{aligned} \tag{2.4.27}$$

Moreover, by (2.4.25),

$$H(x) \ge \int_{(a,x]} \rho(t)\, d\mu(t) \quad \text{and} \quad H(x) \ge \rho(x)\int_{[x,b)} d\mu(t) \quad \text{for all} \quad x \in I. \tag{2.4.28}$$

Together with the inequality $\varepsilon - 1 \le 0$, estimates (2.4.28) imply that

$$\begin{aligned} \text{RHS(2.4.27)} &\le H^{p-\varepsilon}(x)\int_{(a,x]} \Big(\int_{(a,t]} \rho(s)\, d\mu(s)\Big)^{\varepsilon-1} \rho(t)\, d\mu(t) \\ &\quad + \Big(\frac{H}{\rho}\Big)^{p-\varepsilon}(x)\rho^p(x)\int_{[x,b)} \Big(\int_{[t,b)} d\mu(s)\Big)^{\varepsilon-1} d\mu(t). \end{aligned} \tag{2.4.29}$$

Applying Lemmas 1.2.5, 1.2.6 to estimate the integrals on RHS(2.4.29) and using (2.4.28) again, we arrive at

$$\text{(2.4.30)} \quad \begin{aligned} \text{RHS(2.4.29)} &\approx H^{p-\varepsilon}(x)\Big(\int_{(a,x]} \rho(t)\,d\mu(t)\Big)^{\varepsilon} \\ &\quad + \Big(\frac{H}{\rho}\Big)^{p-\varepsilon}(x)\rho^p(x)\Big(\int_{[x,b)} d\mu(t)\Big)^{\varepsilon} \\ &\le H^{p-\varepsilon}(x)H^{\varepsilon}(x) + \Big(\frac{H}{\rho}\Big)^{p-\varepsilon}(x)\rho^p(x)\Big(\frac{H}{\rho}\Big)^{\varepsilon}(x) \\ &\approx H^p(x) \quad \text{for all} \quad x \in I. \end{aligned}$$

Summarizing estimates (2.4.27), (2.4.29) and (2.4.30), we see that

$$\text{(2.4.31)} \qquad \text{RHS(2.4.26)} \lesssim \text{LHS(2.4.26)} \quad \text{for all} \quad x \in I.$$

To prove the reverse estimate, take $x \in I$ and put

$$x_1 = \inf\Big\{t \in (a,x);\; H(t) \ge \frac{1}{3}H(x)\Big\}$$

and

$$x_2 = \sup\Big\{t \in (x,b);\; \Big(\frac{H}{\rho}\Big)(t) \ge \frac{1}{3}\Big(\frac{H}{\rho}\Big)(x)\Big\}$$

(as $H,\ \frac{H}{\rho} \in C(I)$, these two sets are not empty).

Suppose first that $x_1, x_2 \in I$. Then, since $H,\ \frac{H}{\rho} \in C(I)$,

$$\text{(2.4.32)} \qquad H(x_1) = \frac{1}{3}H(x) \quad \text{and} \quad \Big(\frac{H}{\rho}\Big)(x_2) = \frac{1}{3}\Big(\frac{H}{\rho}\Big)(x).$$

As $H \in \mathcal{M}^+(I;\uparrow)$ and $\frac{H}{\rho} \in \mathcal{M}^+(I;\downarrow)$, we obtain from (2.4.32) that

$$\text{(2.4.33)} \qquad \frac{1}{3}H(x) \le H(t) \le H(x) \quad \text{for all} \quad t \in [x_1, x]$$

and

$$\text{(2.4.34)} \qquad \frac{1}{3}\Big(\frac{H}{\rho}\Big)(x) \le \Big(\frac{H}{\rho}\Big)(t) \le \Big(\frac{H}{\rho}\Big)(x) \quad \text{for all} \quad t \in [x, x_2].$$

Inequalities (2.4.33), (2.4.34) and identity (2.4.25) imply that

$$\text{(2.4.35)} \quad \begin{aligned} \text{RHS(2.4.26)} &= \int_{(a,x)} \rho(t)H^{p-1}(t)\,d\mu(t) + \rho^p(x)\int_{[x,b)}\Big(\frac{H}{\rho}\Big)^{p-1}(t)\,d\mu(t) \\ &\ge \int_{[x_1,x)} \rho(t)H^{p-1}(t)\,d\mu(t) + \rho^p(x)\int_{[x,x_2]}\Big(\frac{H}{\rho}\Big)^{p-1}(t)\,d\mu(t) \\ &\gtrsim H^{p-1}(x)\Big[\int_{[x_1,x)} \rho(t)\,d\mu(t) + \rho(x)\int_{[x,x_2]} d\mu(t)\Big] \\ &= H^{p-1}(x)\Big[H(x) - \int_{(a,x_1)} \rho(t)\,d\mu(t) - \rho(x)\int_{(x_2,b)} d\mu(t)\Big]. \end{aligned}$$

Moreover, since by (2.4.25) and (2.4.32),

$$\int_{(a,x_1)} \rho(t)\, d\mu(t) \le H(x_1) = \frac{1}{3}H(x)$$

and

$$\rho(x)\int_{(x_2,b)} d\mu(t) \le \rho(x)\Big(\frac{H}{\rho}\Big)(x_2) = \rho(x)\frac{1}{3}\Big(\frac{H}{\rho}\Big)(x) = \frac{1}{3}H(x), \tag{2.4.36}$$

we see from (2.4.35) that

$$\text{RHS}(2.4.26) \gtrsim \text{LHS}(2.4.26), \tag{2.4.37}$$

which is the desired estimate.

Suppose now that $x_1 = a$ and $x_2 \in I$. Then, instead of the first equality in (2.4.32), we have

$$H(t) \ge \frac{1}{3}H(x) \quad \text{and} \quad t \in (a,x],$$

which implies that (cf. (2.4.33))

$$\frac{1}{3}H(x) \le H(t) \le H(x) \quad \text{for all} \quad t \in (a,x].$$

Proceeding as above, we obtain instead of (2.4.35) that

$$\text{RHS}(2.4.26) \gtrsim H^{p-1}(x)\Big[H(x) - \rho(x)\int_{(x_2,b)} d\mu(t)\Big].$$

Thus, on applying (2.4.36), we arrive at (2.4.37).

When $x_1 \in I$ and $x_2 = b$, or $x_1 = a$ and $x_2 = b$, the proof is analogous. □

Theorem 2.4.4 *Let $I = (a,b) \subseteq \mathbb{R}$, $\rho \in Ads(I)$, $h \in Q_\rho(I)$, $0 \not\equiv h \not\equiv +\infty$, $\alpha, \beta \in [0,+\infty)$, $\mu \in \mathcal{B}^+(I)$, $p \in (0,+\infty)$ and let*

$$h(x) \approx \alpha + \beta\rho(x) + \int_I \min\{\rho(x), \rho(t)\}\, d\mu(t) \quad \textit{for all} \quad x \in I. \tag{2.4.38}$$

Then

$$\Big(\frac{\rho}{h}\Big)^p(x) \approx \alpha_1^p + \beta_1^p\rho^p(x) + V(x) \quad \textit{for all} \quad x \in I, \tag{2.4.39}$$

where

$$V(x) = \int_I \min\{\rho^p(x), \rho^p(t)\} h^{-p-2}(t)\Big(\alpha + \int_{(a,t]} \rho(s)\, d\mu(s)\Big) \times \Big(\beta + \int_{[t,b)} d\mu(s)\Big)\, d\rho(t), \tag{2.4.40}$$

$$\alpha_1 = \lim_{t\to a+}\Big(\frac{\rho}{h}\Big)(t) \quad \textit{and} \quad \beta_1 = \lim_{t \to b-}\Big(\frac{1}{h}\Big)(t). \tag{2.4.41}$$

Proof. First we prove that

$$\alpha_1^p + \beta_1^p \rho^p(x) + V(x) \lesssim \left(\frac{\rho}{h}\right)^p(x) \quad \text{for all} \quad x \in I. \tag{2.4.42}$$

Since $(\rho/h) \in \mathcal{M}^+(I;\uparrow)$,

$$\alpha_1^p = \left[\lim_{t\to a+} \left(\frac{\rho}{h}\right)(t)\right]^p \le \left(\frac{\rho}{h}\right)^p(x), \quad x \in I. \tag{2.4.43}$$

Similarly, $(1/h) \in \mathcal{M}^+(I;\downarrow)$ and so,

$$\beta_1^p \rho^p(x) = \left[\lim_{t\to b-} \left(\frac{1}{h}\right)(t)\right]^p \rho^p(x) \le \left(\frac{\rho}{h}\right)^p(x), \quad x \in I. \tag{2.4.44}$$

Consequently, to verify (2.4.42), it is sufficient to prove that

$$V(x) \lesssim \left(\frac{\rho}{h}\right)^p(x) \quad \text{for all} \quad x \in I. \tag{2.4.45}$$

By (2.4.40), for all $x \in I$,

$$\begin{aligned} V(x) &\le \int_{(a,x]} \rho^{p-1}(t) h^{-p-2}(t) \Big(\alpha + \int_{(a,t]} \rho(s)\, d\mu(s)\Big) \rho(t) \Big(\beta + \int_{[t,b)} d\mu(s)\Big)\, d\rho(t) \\ &+ \rho^p(x) \int_{[x,b)} h^{-p-2}(t) \Big(\alpha + \int_{(a,t]} \rho(s)\, d\mu(s)\Big) \Big(\beta + \int_{[t,b)} d\mu(s)\Big)\, d\rho(t) \\ &=: I_1 + I_2. \end{aligned} \tag{2.4.46}$$

Moreover, (2.4.38) implies that

$$\rho(t)\Big(\beta + \int_{[t,b)} d\mu(s)\Big) \lesssim h(t) \tag{2.4.47}$$

and

$$\alpha + \int_{(a,t]} \rho(s)\, d\mu(s) \lesssim h(t) \tag{2.4.48}$$

for all $t \in I$. Therefore,

$$I_1 \lesssim \int_{(a,x]} \rho^{p-1}(t) h^{-p-1}(t) \Big(\alpha + \int_{(a,t]} \rho(s)\, d\mu(s)\Big)\, d\rho(t) =: \widetilde{I}_1 = \widetilde{I}_1(p,a,x) \tag{2.4.49}$$

and

$$\begin{aligned} I_2 &\lesssim \rho^p(x) \int_{[x,b)} h^{-p-1}(t) \Big(\beta + \int_{[t,b)} d\mu(s)\Big)\, d\rho(t) \\ &=: \rho^p(x) \widetilde{I}_2 = \rho^p(x)\, \widetilde{I}_2(p,x,b). \end{aligned} \tag{2.4.50}$$

Firstly, assume that $(a,b) \subseteq (0,+\infty)$. Let $q \in (0,+\infty)$ and $[x_1,x_2] \subseteq (a,b)$. On applying Lemma 1.2.4[6], we obtain

$$\int_{[x_1,x_2]} d\Big[\Big(\frac{h}{\rho}\Big)^{-q}(t)\Big] = \int_{[x_1,x_2]} (-q)\Big(\frac{h}{\rho}\Big)^{-q-1}(t)\, d\Big[\Big(\frac{h}{\rho}\Big)(t)\Big]. \tag{2.4.51}$$

As

$$\Big(\frac{h}{\rho}\Big)(t) \approx \Big[\frac{1}{\rho(t)}\Big(\alpha + \int_{(a,t]} \rho(s)\, d\mu(s)\Big) + \Big(\beta + \int_{[t,b)} d\mu(s)\Big)\Big], \quad t \in [x_1,x_2],$$

we have

$$\mathrm{RHS}(2.4.51) \approx N_1 + N_2, \tag{2.4.52}$$

where, by Lemma 1.2.8,

$$\begin{aligned} N_2 &:= \int_{[x_1,x_2]} (-q)\Big(\frac{h}{\rho}\Big)^{-q-1}(t)\, d\Big(\beta + \int_{[t,b)} d\mu(s)\Big) \\ &= q\int_{[x_1,x_2]} \Big(\frac{h}{\rho}\Big)^{-q-1}(t)\, d\mu(t) \end{aligned} \tag{2.4.53}$$

and, by Lemma 1.2.2 and Corollary 1.2.9,

$$\begin{aligned} N_1 &:= \int_{[x_1,x_2]} (-q)\Big(\frac{h}{\rho}\Big)^{-q-1}(t)\Big(\alpha + \int_{(a,t]} \rho(s)\, d\mu(s)\Big)\, d\Big[\frac{1}{\rho(t)}\Big] \\ &\quad + \int_{[x_1,x_2]} (-q)\Big(\frac{h}{\rho}\Big)^{-q-1}(t)\frac{1}{\rho(t)}\, d\Big(\alpha + \int_{(a,t]} \rho(s)\, d\mu(s)\Big) \\ &=: N_{11} + N_{12}\,. \end{aligned} \tag{2.4.54}$$

Moreover, by Lemma 1.2.7,

$$N_{12} = \int_{[x_1,x_2]} (-q)\Big(\frac{h}{\rho}\Big)^{-q-1}(t)\, d\mu(t) = -N_2. \tag{2.4.55}$$

Applying Lemma 1.2.4 and Corollary 1.2.9, we obtain (cf. (2.4.49))

$$\begin{aligned} N_{11} &= q\int_{[x_1,x_2]} \Big(\frac{h}{\rho}\Big)^{-q-1}(t)\Big(\alpha + \int_{(a,t]} \rho(s)\, d\mu(s)\Big)\rho^{-2}(t)\, d\rho(t) \\ &=: q\widetilde{I}_1(q,x_1,x_2)\,. \end{aligned} \tag{2.4.56}$$

Consequently (cf. (2.4.54), (2.4.55) and (2.4.56)),

$$N_1 + N_2 = N_{11} + N_{12} + N_2 = N_{11} = q\,\widetilde{I}_1(q,x_1,x_2). \tag{2.4.57}$$

[6]When applying Lemma 1.2.4, we may assume without loss of generality that ρ/h is strictly increasing on (a,b) (otherwise we replace ρ/h with an equivalent function $(\rho/h)\,\omega$, where $\omega : (a,b) \to (1,2)$ is continuous and strictly increasing).

Together with (2.4.51), (2.4.52) and (2.4.56), this yields

$$\left(\frac{h}{\rho}\right)^{-q}(x_2) - \left(\frac{h}{\rho}\right)^{-q}(x_1) \approx q \int_{[x_1,x_2]} \rho^{q-1}(t) h^{-q-1}(t) \Big(\alpha + \int_{(a,t]} \rho(s)\, d\mu(s)\Big)\, d\rho(t). \tag{2.4.58}$$

Using (2.4.58) with $q = p$ and $x_2 = x \in (a,b)$, we see that

$$\begin{aligned} \left(\frac{\rho}{h}\right)^p(x) &\geq \left(\frac{\rho}{h}\right)^p(x) - \alpha_1^p = \lim_{x_1\to a+}\Big[\left(\frac{\rho}{h}\right)^p(x) - \left(\frac{\rho}{h}\right)^p(x_1)\Big] \\ &\approx p \lim_{x_1\to a+} \int_{[x_1,x]} \rho^{p-1}(t) h^{-p-1}(t) \Big(\alpha + \int_{(a,t]} \rho(s)\, d\mu(s)\Big)\, d\rho(t) \\ &= p\, \widetilde{I}_1(p,a,x). \end{aligned} \tag{2.4.59}$$

A similar approach can be applied to estimate $\widetilde{I}_2(p,x,b)$ (cf. (2.4.50)). Indeed, by Lemma 1.2.4 and Corollary 1.2.9,[7]

$$\begin{aligned} -\int_{[x_1,x_2]} d\Big[\left(\frac{1}{h}\right)^q(t)\Big] &\approx -q \int_{[x_1,x_2]} \left(\frac{1}{h}\right)^{q-1}(t)\, d\Big[\left(\frac{1}{h}\right)(t)\Big] \\ &= q \int_{[x_1,x_2]} \left(\frac{1}{h}\right)^{q-1} \frac{1}{h^2(t)}\, dh(t) = q \int_{[x_1,x_2]} \left(\frac{1}{h}\right)^{q+1}(t)\, dh(t). \end{aligned} \tag{2.4.60}$$

Since

$$h(t) \approx \Big(\alpha + \int_{(a,t]} \rho(s)\, d\mu(s)\Big) + \rho(t)\Big(\beta + \int_{[t,b)} d\mu(s)\Big), \quad t \in [x_1, x_2],$$

we have

$$\text{RHS}(2.4.60) \approx M_1 + M_2, \tag{2.4.61}$$

where, by Lemma 1.2.7,

$$\begin{aligned} M_1 &:= q \int_{[x_1,x_2]} \left(\frac{1}{h}\right)^{q+1}(t)\; d\Big(\alpha + \int_{(a,t]} \rho(s)\, d\mu(s)\Big) \\ &= q \int_{[x_1,x_2]} \left(\frac{1}{h}\right)^{q+1}(t) \rho(t)\, d\mu(t), \end{aligned} \tag{2.4.62}$$

and, by Lemma 1.2.3 and Corollary 1.2.9,

$$\begin{aligned} M_2 &:= q \int_{[x_1,x_2]} \left(\frac{1}{h}\right)^{q+1}(t) \rho(t)\; d\Big(\beta + \int_{[t,b)} d\mu(s)\Big) \\ &\quad + q \int_{[x_1,x_2]} \left(\frac{1}{h}\right)^{q+1}(t) \Big(\beta + \int_{[t,b)} d\mu(s)\Big)\, d\rho(t) \\ &=: M_{21} + M_{22}\,. \end{aligned} \tag{2.4.63}$$

[7] Again, when using Lemma 1.2.4, we replace $(1/h)$ by an equivalent function which is strictly increasing on (a,b).

Moreover, by Lemma 1.2.8, (2.4.63) and (2.4.62),

$$M_{21} = -q \int_{[x_1,x_2]} \left(\frac{1}{h}\right)^{q+1}(t)\rho(t)\,d\mu(t) = -M_1. \tag{2.4.64}$$

Furthermore,

$$M_{22} = q \int_{[x_1,x_2]} h^{-q-1}(t)\Big(\beta + \int_{[t,b)} d\mu(s)\Big)\,d\rho(t) =: q\,\widetilde{I}_2(q,x_1,x_2) \tag{2.4.65}$$

(cf. (2.4.50)). Consequently (cf. (2.4.63)–(2.4.65)),

$$M_1 + M_2 = M_1 + M_{21} + M_{22} = M_{22} = q\,\widetilde{I}_2(q,x_1,x_2).$$

Together with (2.4.60), (2.4.61) and (2.4.50), this yields

$$\left(\frac{1}{h}\right)^{q}(x_1) - \left(\frac{1}{h}\right)^{q}(x_2) \approx q \int_{[x_1,x_2]} h^{-q-1}(t)\Big(\beta + \int_{[t,b)} d\mu(s)\Big)\,d\rho(t). \tag{2.4.66}$$

Using (2.4.66) with $q = p$ and $x_1 = x \in (a,b)$, we obtain that

$$\begin{aligned}\left(\frac{1}{h}\right)^{p}(x) \geq \left(\frac{1}{h}\right)^{p}(x) - \beta_1^p &= \lim_{x_2\to b-}\left[\left(\frac{1}{h}\right)^{p}(x) - \left(\frac{1}{h}\right)^{p}(x_2)\right] \\ &\approx p \lim_{x_2\to b-}\int_{[x,x_2]} h^{-p-1}(t)\Big(\beta + \int_{[t,b)} d\mu(s)\Big)\,d\rho(t) \\ &= p\,\widetilde{I}_2(p,x,b).\end{aligned} \tag{2.4.67}$$

Now, by (2.4.46), (2.4.49), (2.4.50), (2.4.59) and (2.4.67), we arrive at

$$\begin{aligned}V(x) \leq I_1 + I_2 &\lesssim \widetilde{I}_1(p,a,x) + \rho^p(x)\,\widetilde{I}_2(p,x,b) \\ &\lesssim \left(\frac{\rho}{h}\right)^{p}(x) + \rho^p(x)\left(\frac{1}{h}\right)^{p}(x) \\ &\approx \left(\frac{\rho}{h}\right)^{p}(x)\end{aligned}$$

for all $x \in (a,b)$, which is the desired estimate (2.4.45).

If $(a,b) \not\subseteq (0,+\infty)$, one can still obtain (2.4.45) by applying the method used above to the function

$$\left(\frac{t}{h(\rho^{-1}(t))}\right)^{p}, \quad t \in (\rho(a+),\rho(b-)) \subseteq (0,+\infty), \tag{2.4.68}$$

(rather than to ρ/h) and then passing to the original problem by changing the variables.

Now, we are going to prove the reverse estimate to (2.4.42), that is,

$$\alpha_1^p + \beta_1^p\rho^p(x) + V(x) \gtrsim \left(\frac{\rho}{h}\right)^{p}(x) \quad \text{for all} \quad x \in I. \tag{2.4.69}$$

Take $x \in (a,b)$. Assume first that

$$\left(\frac{\rho}{h}\right)^p(x) \le 2^{p/(p+1)}\alpha_1^p \quad \text{or} \quad \left(\frac{\rho}{h}\right)^p(x) \le 2^{p/(p+1)}\beta_1^p\rho^p(x). \tag{2.4.70}$$

Then it is plain that (2.4.69) holds. Suppose now that (2.4.70) is not satisfied, that is,

$$\left(\frac{\rho}{h}\right)^p(x) > 2^{p/(p+1)}\alpha_1^p \quad \text{and} \quad \left(\frac{\rho}{h}\right)^p(x) > 2^{p/(p+1)}\beta_1^p\rho^p(x). \tag{2.4.71}$$

Assume also that $(a,b) \subseteq (0,+\infty)$ (if it is not the case, one should again work with the function (2.4.68) rather than with ρ/h). Moreover, if $a < x_1 < x_2 < b$ and $x \in (x_1, x_2)$, then by (2.4.40),

$$\begin{aligned} V(x) & \\ &\gtrsim \Big[\int_{[x_1,x]} \rho^p(t)h^{-p-2}(t)\Big(\alpha+\int_{(a,t]}\rho(s)\,d\mu(s)\Big)\,d\rho(t)\Big]\Big(\beta+\int_{[x,b)}d\mu(s)\Big) \\ &\quad +\rho^p(x)\Big[\int_{[x,x_2]} h^{-p-2}(t)\Big(\beta+\int_{[t,b)}d\mu(s)\Big)\,d\rho(t)\Big]\Big(\alpha+\int_{(a,x]}\rho(s)d\mu(s)\Big) \\ &=: \widetilde{I}_1(p+1,x_1,x)\Big(\beta+\int_{[x,b)}d\mu(s)\Big) \\ &\quad + \rho^p(x)\Big(\alpha+\int_{(a,x]}\rho(s)\,d\mu(s)\Big)\widetilde{I}_2(p+1,x,x_2) \end{aligned} \tag{2.4.72}$$

(for $\widetilde{I}_1$ and $\widetilde{I}_2$ we refer to (2.4.56) and (2.4.65), respectively). Put $q = p+1$. Then, using (2.4.58) and (2.4.56) with x instead of x_2, we arrive at

$$\widetilde{I}_1(p+1,x_1,x) \approx \left(\frac{\rho}{h}\right)^{p+1}(x) - \left(\frac{\rho}{h}\right)^{p+1}(x_1). \tag{2.4.73}$$

Similarly, cf. (2.4.66) and (2.4.65), we obtain

$$\widetilde{I}_2(p+1,x,x_2) \approx \left[\left(\frac{1}{h}\right)^{p+1}(x) - \left(\frac{1}{h}\right)^{p+1}(x_2)\right]. \tag{2.4.74}$$

Therefore, (2.4.72)–(2.4.74) imply that

$$\begin{aligned} V(x) &\gtrsim \Big(\beta+\int_{[x,b)}d\mu(s)\Big)\lim_{x_1\to a+}\left[\left(\frac{\rho}{h}\right)^{p+1}(x) - \left(\frac{\rho}{h}\right)^{p+1}(x_1)\right] \\ &\quad + \rho^p(x)\Big(\alpha+\int_{(a,x]}\rho(s)\,d\mu(s)\Big)\lim_{x_2\to b-}\left[\left(\frac{1}{h}\right)^{p+1}(x) - \left(\frac{1}{h}\right)^{p+1}(x_2)\right] \\ &= \Big(\beta+\int_{[x,b)}d\mu(s)\Big)\left[\left(\frac{\rho}{h}\right)^{p+1}(x) - \alpha_1^{p+1}\right] \\ &\quad + \rho^p(x)\Big(\alpha+\int_{(a,x]}\rho(s)\,d\mu(s)\Big)\left[\left(\frac{1}{h}\right)^{p+1}(x) - \beta_1^{p+1}\right]. \end{aligned}$$

Together with (2.4.71) and (2.4.38), this yields

$$
\begin{aligned}
V(x) &\gtrsim \Big(\beta + \int_{[x,b)} d\mu(s)\Big)\Big(\frac{\rho}{h}\Big)^{p+1}(x) + \rho^p(x)\Big(\alpha + \int_{(a,x]} \rho(s)\, d\mu(s)\Big)\Big(\frac{1}{h}\Big)^{p+1}(x) \\
&= \Big(\frac{\rho}{h}\Big)^p(x)\frac{1}{h(x)}\Big[\rho(x)\Big(\beta + \int_{[x,b)} d\mu(s)\Big) + \Big(\alpha + \int_{(a,x]} \rho(s)\, d\mu(s)\Big)\Big] \\
&\approx \Big(\frac{\rho}{h}\Big)^p(x),
\end{aligned}
$$

which implies that (2.4.69) holds. □

The next assertion shows that if a function h is equivalent on I to the ρ-fundamental function of the space $L^1(w, I, \mu)$, then, for any $p \in (0, +\infty]$, there is a weight W_p on I such that h is also equivalent on I to the ρ-fundamental function of the space $L^p(W_p, I, \mu)$.

Lemma 2.4.5 *Let $I = (a, b) \subseteq \mathbb{R}$, $\rho \in Ads(I)$, $w \in \mathcal{W}(I)$, $\mu \in \mathcal{B}^+(I)$ and $p \in (0, +\infty]$. Assume that*

$$
h(x) \approx H(x) := \|\min\{\rho(\cdot), \rho(x)\}\|_{1,w,I,\mu} \quad \text{for all } x \in I. \tag{2.4.75}
$$

Then

$$
h(x) \approx \|\min\{\rho(\cdot), \rho(x)\}\|_{p,W_p,I,\mu} \quad \text{for all } x \in I, \tag{2.4.76}
$$

where

$$
W_p(x) = \Big(\frac{H}{\rho}\Big)(x)\Big(\frac{\rho w}{H}\Big)^{1/p}(x), \quad x \in I. \tag{2.4.77}
$$

Proof. If $p \in (0, +\infty)$, then the result is a consequence of Theorem 2.4.3. Thus, it is sufficient to verify that

$$
h(x) \approx \|\min\{\rho(\cdot), \rho(x)\}\|_{\infty,W_\infty,I,\mu} \quad \text{for all } x \in I. \tag{2.4.78}
$$

Using the facts that $H \in \mathcal{M}^+(I;\uparrow)$, $(H/\rho) \in \mathcal{M}^+(I;\downarrow)$ and $W_\infty = H/\rho$, we obtain, for all $x \in I$,

$$
\begin{aligned}
\mathrm{RHS}(2.4.78) &= \|\min\{\rho(\cdot), \rho(x)\}W_\infty(\cdot)\|_{\infty,I,\mu} \\
&= \max\{\|\min\{\rho(\cdot), \rho(x)\}W_\infty(\cdot)\|_{\infty,(a,x],\mu}, \\
&\qquad\quad \|\min\{\rho(\cdot), \rho(x)\}W_\infty(\cdot)\|_{\infty,(x,b),\mu}\} \\
&= \max\{\|\rho(\cdot)W_\infty(\cdot)\|_{\infty,(a,x],\mu}, \ \|\rho(x)W_\infty(\cdot)\|_{\infty,(x,b),\mu}\} \\
&= \max\{\|H(\cdot)\|_{\infty,(a,x],\mu}, \ \rho(x)\|(H/\rho)(\cdot)\|_{\infty,(x,b),\mu}\} \\
&= H(x).
\end{aligned}
$$

Together with (2.4.75), this yields (2.4.78). □

We shall also need the following lemma.

Lemma 2.4.6 *Let $I = (a,b) \subseteq \mathbb{R}$, $\rho \in Ads(I)$, $w \in \mathcal{W}(I)$ and $\mu \in \mathcal{B}^+(I)$. Assume that*

$$h(x) \approx H(x) := \|\min\{\rho(\cdot), \rho(x)\}\|_{1,w,I,\mu} \quad \textit{for all } x \in I. \tag{2.4.79}$$

Then

$$\left(\frac{\rho}{h}\right)(x) \approx \|\min\{\rho(\cdot), \rho(x)\}\|_{\infty,1/H,I,\mu} \quad \textit{for all } x \in I. \tag{2.4.80}$$

Proof. Using the facts that $(\rho/H) \in \mathcal{M}^+(I;\uparrow)$ and $(1/H) \in \mathcal{M}^+(I;\downarrow)$, we obtain

$$\begin{aligned}
\mathrm{RHS}(2.4.80) &= \|\min\{\rho(\cdot), \rho(x)\}/H(\cdot)\|_{\infty,I,\mu} \\
&= \max\{\|\min\{\rho(\cdot), \rho(x)\}/H(\cdot)\|_{\infty,(a,x],\mu}, \\
&\qquad\qquad \|\min\{\rho(\cdot), \rho(x)\}/H(\cdot)\|_{\infty,(x,b),\mu}\} \\
&= \max\{\|(\rho/H)(\cdot)\|_{\infty,(a,x],\mu},\ \rho(x)\|(1/H)(\cdot)\|_{\infty,(x,b),\mu}\} \\
&= (\rho/H)(x) \quad \text{for all } x \in I.
\end{aligned}$$

Since, by (2.4.79), $h \approx H$ on I, we see that (2.4.80) is satisfied. □

Chapter 3

ρ-discretizing sequences

3.1 Motivation

We begin by describing a construction which motivates the definition in the next chapter. To any function $h \in Q_\rho(I)$, $0 \not\equiv h \not\equiv +\infty$, we shall assign an increasing sequence $\{t_j\}_{j=J_-}^{J_+} \subset I := (a,b) \subseteq \mathbb{R}$ which will be used to define a convenient decomposition of the interval I. This sequence is constructed by means of the following *algorithm*.

Choose $\alpha \in (1,+\infty)$ and $t_0 \in I$. Then $t_1 < t_2 < \dots$ are defined inductively by the requirement that $t_j > t_{j-1}$ $(j = 1,2,\dots)$ is the smallest number in I for which both

$$\alpha h(t_{j-1}) \le h(t_j) \tag{3.1.1}$$

and

$$\alpha\Big(\frac{\rho}{h}\Big)(t_{j-1}) \le \Big(\frac{\rho}{h}\Big)(t_j) \tag{3.1.2}$$

hold. Either the inductive procedure of selecting t_j never terminates and then we put $J_+ = +\infty$, or this inductive procedure terminates at the j-th step (say $j = J_+ \in \mathbb{N}$) if it is impossible to find $t_{j+1} \in (t_j, b)$ satisfying both (3.1.1) and (3.1.2) (with $j = J_+ + 1$).

To construct the points $t_{-1} > t_{-2} > \dots$, we can use the preceding procedure of constructing the sequence $t_1 < t_2 < \dots$ but now applied to $\tilde{t}_0 := t_0$ and to the functions $\tilde{\rho}$ and $\tilde{h}$ given by

$$\tilde{\rho} := \frac{1}{\rho \circ g} \quad \text{and} \quad \tilde{h} = \frac{1}{h \circ g}, \tag{3.1.3}$$

where g is a continuous, decreasing bijection of the interval $I = (a,b)$ onto itself satisfying

$$g(a) = b, \quad g(t_0) = t_0 \quad \text{and} \quad g(b) = a.$$

Then, having obtained the sequence $\tilde{t}_1 < \tilde{t}_2 < \dots$, we put

$$t_{-j} = g(\tilde{t}_j), \quad j = 1, 2, \dots. \tag{3.1.4}$$

If the procedure of defining $\tilde{t}_j$ terminates at the $\tilde{J}_+$th step, say, we put

$$J_- = -\tilde{J}_+. \tag{3.1.5}$$

Otherwise we put $J_- = -\infty$.

In the next lemma we summarize some of the properties of the sequence $\{t_j\}_{j=J_-}^{J_+}$ given by the algorithm mentioned above. (Recall that we make use of Convention 1.1.1 (iv).)

Lemma 3.1.1 *Let $I = (a, b) \subseteq \mathbb{R}$, $\rho \in Ads(I)$, $h \in Q_\rho(I)$, $0 \not\equiv h \not\equiv +\infty$, and $\alpha \in (1, +\infty)$. Let $\{t_j\}_{j=J_-}^{J_+} \subset I$ be the increasing sequence defined above. Then:*

(i) *The inequalities*

$$\alpha h(t_{j-1}) \le h(t_j) \quad \text{and} \quad \alpha\Big(\frac{\rho}{h}\Big)(t_{j-1}) \le \Big(\frac{\rho}{h}\Big)(t_j) \tag{3.1.6}$$

hold for all $j \in \mathbb{Z}$ satisfying $J_- < j \le J_+$.

(ii) *$0 \le J_+ \le +\infty$, and $J_+ = +\infty$ if and only if*

$$\lim_{t \to b-} h(t) = +\infty \quad \text{and} \quad \lim_{t \to b-} \Big(\frac{\rho}{h}\Big)(t) = +\infty. \tag{3.1.7}$$

(iii) *If $J_+ = +\infty$, then*

$$\lim_{j \to +\infty} t_j = b. \tag{3.1.8}$$

(iv) *$-\infty \le J_- \le 0$, and $J_- = -\infty$ if and only if*

$$\lim_{t \to a+} h(t) = 0 \quad \text{and} \quad \lim_{t \to a+} \Big(\frac{\rho}{h}\Big)(t) = 0. \tag{3.1.9}$$

(v) *If $J_- = -\infty$, then*

$$\lim_{j \to -\infty} t_j = a. \tag{3.1.10}$$

(vi) *There is a positive constant C such that:*
For all $j \in \mathbb{Z} \cap (J_-, J_+]$,

$$h(t_j) \le C h(t_{j-1}) \tag{3.1.11}$$

or

$$\Big(\frac{\rho}{h}\Big)(t_j) \le C\Big(\frac{\rho}{h}\Big)(t_{j-1}). \tag{3.1.12}$$

If $J_+ < +\infty$, then

$$h(b) \le C h(t_{J_+}) \tag{3.1.13}$$

or

$$\left(\frac{\rho}{h}\right)(b) \le C\left(\frac{\rho}{h}\right)(t_{J_+}). \tag{3.1.14}$$

If $J_- > -\infty$, *then*

$$h(t_{J_-}) \le Ch(a) \tag{3.1.15}$$

or

$$\left(\frac{\rho}{h}\right)(t_{J_-}) \le C\left(\frac{\rho}{h}\right)(a). \tag{3.1.16}$$

Proof. (i) The inequalities (3.1.6) are clear if $j \in \mathbb{N}$ and $1 \le j \le J_+$. On the other hand, if $\tilde{h}$ is given by (3.1.3), then

$$\alpha\tilde{h}(\tilde{t}_{j-1}) \le \tilde{h}(\tilde{t}_j), \quad j \in \mathbb{N},\ 1 \le j \le \widetilde{J}_+. \tag{3.1.17}$$

Using (3.1.3) and (3.1.4), we can rewrite (3.1.17) as

$$\alpha\frac{1}{h(t_{1-j})} \le \frac{1}{h(t_{-j})}, \quad j \in \mathbb{N},\ 1 \le j \le \widetilde{J}_+,$$

which, together with (3.1.5), implies that the first condition in (3.1.6) holds for all $j \in \mathbb{N}$ satisfying $J_- < j \le 0$. The second condition in (3.1.6) can be verified analogously for such j's.

(ii) Firstly, it is clear that $J_+ = +\infty$ if (3.1.7) holds. Indeed, since, by (3.1.7), in any step of our algorithm there are points $t \in I$ such that $t_{j-1} < t$ and both

$$\alpha h(t_{j-1}) \le h(t) \tag{3.1.18}$$

and

$$\alpha\left(\frac{\rho}{h}\right)(t_{j-1}) \le \left(\frac{\rho}{h}\right)(t) \tag{3.1.19}$$

hold. Consequently, since $h, \frac{\rho}{h} \in C(I)$ (cf. (2.1.6)), the set of values of t satisfying (3.1.18) and (3.1.19) has a minimum, t_j say; note that this smallest t $(= t_j)$ turns at least one of the inequalities (3.1.18), (3.1.19) into the equality. This implies that $J_+ = +\infty$.

Secondly, assume that $J_+ = +\infty$. Then, by (3.1.6),

$$\alpha^j h(t_0) \le \alpha^{j-1}h(t_1) \le \cdots \le \alpha^2 h(t_{j-2}) \le \alpha h(t_{j-1}) \le h(t_j) \tag{3.1.20}$$

for any $j = 1, 2, \cdots$, which shows that $h(t_j) \to +\infty$ as $j \to +\infty$. Moreover, since $t_1 < t_2 < \ldots$ and $h \in \mathcal{M}^+(I;\uparrow)$, it follows that

$$\lim_{t\to b-} h(t) = +\infty,$$

which is the first condition in (3.1.7). The second one can be verified similarly.

(iii) Let $J_+ = +\infty$. Suppose that

$$\lim_{j\to+\infty} t_j = c \in (a,b). \tag{3.1.21}$$

Since $h \in C(I)$, we have from (3.1.21) that

$$\lim_{j\to+\infty} h(t_j) = h(c) < +\infty$$

(cf. (2.1.6)). On the other hand, (3.1.20) shows that

$$\lim_{j\to+\infty} h(t_j) \geq \lim_{j\to+\infty} \alpha^j h(t_0) = +\infty,$$

which is a contradiction. Together with the fact that the sequence $\{t_j\} \subset I$ is (strictly) increasing, this implies (3.1.8).

The proofs of (iv) and (v) are analogous to those of (ii) and (iii), respectively.

(We can even write that (iv) is a consequence of (ii) and our construction; similarly for (v).)

(vi) Assume that $j \in \mathbb{Z} \cap (J_-, J_+]$. Since our algorithm has not terminated at the j-th step, at least one of the conditions

$$\alpha h(t_{j-1}) = h(t_j) \tag{3.1.22}$$

or

$$\alpha\Big(\frac{\rho}{h}\Big)(t_{j-1}) = \Big(\frac{\rho}{h}\Big)(t_j) \tag{3.1.23}$$

holds. If (3.1.22) or (3.1.23) is satisfied, then (3.1.11) or (3.1.12), respectively, follows with $C = \alpha$.

If $J_+ < +\infty$, then, by definition of t_{J_+}, we know that

$$\alpha h(t_{J_+}) > h(t) \quad \text{for all} \quad t \in (t_{J_+}, b) \tag{3.1.24}$$

or

$$\alpha\Big(\frac{\rho}{h}\Big)(t_{J_+}) > \Big(\frac{\rho}{h}\Big)(t) \quad \text{for all} \quad t \in (t_{J_+}, b). \tag{3.1.25}$$

If (3.1.24) or (3.1.25) is satisfied, then

$$h(b) = \lim_{t\to b-} h(t) \leq \alpha h(t_{J_+})$$

or

$$\Big(\frac{\rho}{h}\Big)(b) = \lim_{t\to b-} \Big(\frac{\rho}{h}\Big)(t) \leq \alpha\Big(\frac{\rho}{h}\Big)(t_{J_+})$$

and (3.1.13) or (3.1.14), respectively, follows with $C = \alpha$.

If $J_- > -\infty$, then, by definition of t_{J_-}, we know that

$$\alpha h(t) > h(t_{J_-}) \quad \text{for all } t \in (a, t_{J_-}) \tag{3.1.26}$$

or

$$\alpha\Big(\frac{\rho}{h}\Big)(t) > \Big(\frac{\rho}{h}\Big)(t_{J_-}) \quad \text{for all } t \in (a, t_{J_-}). \tag{3.1.27}$$

Assuming (3.1.26) or (3.1.27), we arrive at

$$\alpha h(a) = \lim_{t\to a+} \alpha h(t) \geq h(t_{J_-})$$

or

$$\alpha\Big(\frac{\rho}{h}\Big)(a) = \lim_{t\to a+} \alpha\Big(\frac{\rho}{h}\Big)(t) \geq \Big(\frac{\rho}{h}\Big)(t_{J_-}),$$

and (3.1.15) or (3.1.16), respectively, follows $C = \alpha$.

□

3.2 The set $DS(h,\rho,I)$

Motivated by Lemma 3.1.1, we arrive at the following definition.

Definition 3.2.1 Let $I = (a,b) \subseteq \mathbb{R}$, $\rho \in Ads(I)$ and $h \in Q_\rho(I)$. A strictly increasing sequence $\{t_j\}_{j=J_-}^{J_+} \subset I$, where $-\infty \leq J_- \leq 0 \leq J_+ \leq +\infty$, is said to be a *$\rho$-discretizing sequence* of h if:

(i) There is $\alpha \in (1,+\infty)$ such that the inequalities

$$\alpha h(t_{j-1}) \leq h(t_j) \quad \text{and} \quad \alpha\Big(\frac{\rho}{h}\Big)(t_{j-1}) \leq \Big(\frac{\rho}{h}\Big)(t_j) \tag{3.2.1}$$

hold for all $j \in \mathbb{Z}$ satisfying $J_- < j \leq J_+$.

(ii) There is a positive constant C such that:

For any $j \in \mathbb{Z} \cap (J_-, J_+]$,

$$h(t_j) \leq Ch(t_{j-1}) \tag{3.2.2}$$

or

$$\Big(\frac{\rho}{h}\Big)(t_j) \leq C\Big(\frac{\rho}{h}\Big)(t_{j-1}). \tag{3.2.3}$$

If $J_+ < +\infty$, then

$$h(b) \leq Ch(t_{J_+}) \tag{3.2.4}$$

or

$$\Big(\frac{\rho}{h}\Big)(b) \leq C\Big(\frac{\rho}{h}\Big)(t_{J_+}). \tag{3.2.5}$$

If $J_- > -\infty$, then

$$h(t_{J_-}) \leq Ch(a) \tag{3.2.6}$$

or

$$(3.2.7) \qquad \left(\frac{\rho}{h}\right)(t_{J_-}) \le C\left(\frac{\rho}{h}\right)(a).$$

The set of all ρ-discretizing sequences of the function $h \in Q_\rho(I)$ will be denoted by $DS(h,\rho,I)$. The subset of those elements of $DS(h,\rho,I)$ which satisfy (3.2.1) with a given $\alpha \in (1,+\infty)$ will be denoted by $DS(h,\rho,I,\alpha)$. The elements of $DS(h,\rho,I,\alpha)$ will be called (ρ,α)-*discretizing sequences* of $h \in Q_\rho(I)$.

Remark 3.2.2 Note that the condition (ii) of Definition 3.2.1 is equivalent to the following one.

$\overline{\text{(ii)}}$ There is a positive constant C such that:

For any $j \in \mathbb{Z} \cap (J_-, J_+]$,

$$(3.2.8) \qquad C^{-1}h(t_j) \le h(t) \le h(t_j) \quad \text{for all } t \in [t_{j-1}, t_j]$$

or

$$(3.2.9) \qquad C^{-1}\left(\frac{\rho}{h}\right)(t_j) \le \left(\frac{\rho}{h}\right)(t) \le \left(\frac{\rho}{h}\right)(t_j) \quad \text{for all } t \in [t_{j-1}, t_j].$$

If $J_+ < +\infty$, then

$$(3.2.10) \qquad h(t_{J_+}) \le h(t) \le C\,h(t_{J_+}) \quad \text{for all } t \in [t_{J_+}, b)$$

or

$$(3.2.11) \qquad \left(\frac{\rho}{h}\right)(t_{J_+}) \le \left(\frac{\rho}{h}\right)(t) \le C\left(\frac{\rho}{h}\right)(t_{J_+}) \quad \text{for all } t \in [t_{J_+}, b).$$

If $J_- > -\infty$, then

$$(3.2.12) \qquad C^{-1}h(t_{J_-}) \le h(t) \le h(t_{J_-}) \quad \text{for all } t \in (a, t_{J_-}]$$

or

$$(3.2.13) \qquad C^{-1}\left(\frac{\rho}{h}\right)(t_{J_-}) \le \left(\frac{\rho}{h}\right)(t) \le \left(\frac{\rho}{h}\right)(t_{J_-}) \quad \text{for all } t \in (a, t_{J_-}].$$

Indeed, it is clear that condition $\overline{\text{(ii)}}$ implies condition (ii) of Definition 3.2.1. On the other hand, using the monotonicity of h and $\frac{\rho}{h}$, we obtain that, for any $j \in \mathbb{Z} \cap (J_-, J_+]$,

$$h(t_{j-1}) \le h(t) \le h(t_j) \quad \text{for all } t \in [t_{j-1}, t_j]$$

and

$$\left(\frac{\rho}{h}\right)(t_{j-1}) \le \left(\frac{\rho}{h}\right)(t) \le \left(\frac{\rho}{h}\right)(t_j) \quad \text{for all } t \in [t_{j-1}, t_j].$$

Since also, by (3.2.2) or (3.2.3),

$$h(t_{j-1}) \ge C^{-1}h(t_j)$$

or

$$\Big(\frac{\rho}{h}\Big)(t_{j-1}) \geq C^{-1}\Big(\frac{\rho}{h}\Big)(t_j),$$

we see that

$$C^{-1}h(t_j) \leq h(t_{j-1}) \leq h(t) \leq h(t_j) \quad \text{for all } t \in [t_{j-1}, t_j]$$

or

$$C^{-1}\Big(\frac{\rho}{h}\Big)(t_j) \leq \Big(\frac{\rho}{h}\Big)(t_{j-1}) \leq \Big(\frac{\rho}{h}\Big)(t) \leq \Big(\frac{\rho}{h}\Big)(t_j) \quad \text{for all } t \in [t_{j-1}, t_j],$$

and (3.2.8) or (3.2.9), respectively, follows. The rest of the proof that condition (ii) of Definition 3.2.1 implies condition $\overline{\text{(ii)}}$ can be done analogously.

Remarks 3.2.3 (i) As a consequence of Lemma 3.1.1, we have that $DS(h,\rho,I,\alpha) \neq \emptyset$ for any $h \in Q_\rho(I)$.

(ii) Let $h \in Q_\rho(I)$. Then, by Lemma 2.1.3, $\frac{\rho}{h} \in Q_\rho(I)$ and also $h^p \in Q_{\rho^p}(I)$ for any $p \in (0,+\infty)$. Moreover, one can easily verify that the following implications hold:

$$\{t_j\}_{j=J_-}^{J_+} \in DS(h,\rho,I,\alpha) \Rightarrow \{t_j\}_{j=J_-}^{J_+} \in DS\Big(\frac{\rho}{h},\rho,I,\alpha\Big),$$
$$\{t_j\}_{j=J_-}^{J_+} \in DS(h,\rho,I,\alpha) \Rightarrow \{t_j\}_{j=J_-}^{J_+} \in DS(h^p,\rho^p,I,\alpha^p),\ p \in (0,+\infty).$$

In the next two lemmas we mention some more properties of ρ-discretizing sequences of $h \in Q_\rho(I)$.

Lemma 3.2.4 *Let $I=(a,b) \subseteq \mathbb{R}$, $h \in Q_\rho(I)$, $0 \not\equiv h \not\equiv +\infty$, $\alpha \in (1,+\infty)$ and let $\{t_j\}_{j=J_-}^{J_+} \in DS(h,\rho,I,\alpha)$.*

(i) *Then the inequalities*

$$\alpha\rho(t_{j-1}) \leq \rho(t_j) \tag{3.2.14}$$

hold for all $j \in \mathbb{Z}$ satisfying $J_- < j \leq J_+$.

(ii) *If $J_+ = +\infty$, then*

$$\lim_{j\to+\infty} t_j = b. \tag{3.2.15}$$

(iii) *$J_+ = +\infty$ if and only if*

$$\lim_{t\to b-} h(t) = +\infty \quad \text{and} \quad \lim_{t\to b-}\Big(\frac{\rho}{h}\Big)(t) = +\infty. \tag{3.2.16}$$

(iv) *If $J_- = -\infty$, then*

$$\lim_{j\to-\infty} t_j = a. \tag{3.2.17}$$

(v) *$J_- = -\infty$ if and only if*

$$\lim_{t\to a+} h(t) = 0 \quad \text{and} \quad \lim_{t\to a+}\Big(\frac{\rho}{h}\Big)(t) = 0. \tag{3.2.18}$$

Proof. (i) By the second inequality in (3.2.1) and the monotonicity of h,

$$\left(\frac{\rho}{h}\right)(t_j) \geq \alpha\left(\frac{\rho}{h}\right)(t_{j-1}) \geq \alpha\frac{\rho(t_{j-1})}{h(t_j)}$$

for all $j \in \mathbb{Z}$ satisfying $J_- < j \leq J_+$, which implies (3.2.14).

(ii) If $J_+ = +\infty$, then (3.2.1) implies that, for all $j \in \mathbb{N}$,

$$\alpha^j h(t_0) \leq \alpha^{j-1}h(t_1) \leq \alpha^{j-2}h(t_2) \leq \cdots \leq \alpha h(t_{j-1}) \leq h(t_j)$$

and

$$\alpha^j\left(\frac{\rho}{h}\right)(t_0) \leq \alpha^{j-1}\left(\frac{\rho}{h}\right)(t_1) \leq \alpha^{j-2}\left(\frac{\rho}{h}\right)(t_2) \leq \cdots \leq \alpha\left(\frac{\rho}{h}\right)(t_{j-1}) \leq \left(\frac{\rho}{h}\right)(t_j).$$

Consequently,

$$\lim_{j\to+\infty} h(t_j) = +\infty \quad \text{and} \quad \lim_{j\to+\infty}\left(\frac{\rho}{h}\right)(t_j) = +\infty. \tag{3.2.19}$$

Assuming that $\lim_{j\to+\infty} t_j = c \in (a,b)$ and using also the inclusion $h, \frac{\rho}{h} \in C(I)$, we obtain

$$\lim_{j\to+\infty} h(t_j) = h(c) \in \mathbb{R} \quad \text{and} \quad \lim_{j\to+\infty}\left(\frac{\rho}{h}\right)(t_j) = \left(\frac{\rho}{h}\right)(c) \in \mathbb{R},$$

which contradicts (3.2.19). Together with the fact that the sequence $\{t_j\}_{j=J_-}^{J_+} \subset I$ is strictly increasing, this implies (3.2.15).

(iii) If $J = +\infty$, then (3.2.19) is satisfied. Moreover, by (ii), (3.2.15) holds. Therefore, (3.2.16) is a consequence of the fact that

$$h, \ \frac{\rho}{h} \in \mathcal{M}^+(I;\uparrow). \tag{3.2.20}$$

Assume now that (3.2.16) is satisfied. To prove that $J_+ = +\infty$, suppose the opposite, that is, $J_+ < +\infty$. Then $t_{J_+} \in (a,b)$ and, by Remark 3.2.2, estimate (3.2.10) or (3.2.11) holds. Since $h(t_{J_+})$, $(\frac{\rho}{h})(t_{J_+}) \in \mathbb{R}$, this contradicts (3.2.16). Consequently, $J_+ = +\infty$.

The proofs of (iv) and (v) are analogous to those of (ii) and (iii), respectively. □

Lemma 3.2.5 *Let $I = (a,b) \subseteq \mathbb{R}$, $h \in Q_\rho(I)$, $\alpha \in (1,+\infty)$ and let $\{t_j\}_{j=J_-}^{J_+} \in DS(h,\rho,I,\alpha)$. Put $\mathcal{S} = \{j \in \mathbb{Z};\ J_- < j \leq J_+\}$. Then there is a decomposition*

$$\mathcal{S} = \mathcal{S}_1 \cup \mathcal{S}_2, \quad \mathcal{S}_1 \cap \mathcal{S}_2 = \emptyset,$$

such that

$$h(t) \approx h(t_j) \quad \text{for all } t \in [t_{j-1}, t_j] \text{ and every } j \in \mathcal{S}_1 \tag{3.2.21}$$

and

$$\left(\frac{\rho}{h}\right)(t) \approx \left(\frac{\rho}{h}\right)(t_j) \quad \text{for all } t \in [t_{j-1}, t_j] \text{ and every } j \in \mathcal{S}_2. \tag{3.2.22}$$

Proof. Defining $\mathcal{S}_1$ as the collection of all integers $j \in \mathcal{S}$ for which (3.2.8) holds and putting $\mathcal{S}_2 = \mathcal{S} \setminus \mathcal{S}_1$, we arrive at the desired result. □

3.3 The set $CS(h, \rho, I)$

Now, let $h \in Q_\rho(I)$, $I = (a, b)$. We are going to assign to any ρ-discretizing sequence $\{t_j\}_{j=J_-}^{J_+}$ of h another strictly increasing sequence, say $\{x_k\}_{k=K_-}^{K_+}$, which we shall call the *ρ-covering sequence* of h and which is defined as follows:

(i) If $J_+ < +\infty$, we put $K_+ = J_+ + 1$ and $x_{K_+} = b$.
(ii) If $J_- > -\infty$, we put $K_- = J_- - 1$ and $x_{K_-} = a$.
(iii) If $J_+ = +\infty$, we put $K_+ = J_+$.
(iv) If $J_- = -\infty$, we put $K_- = J_-$.
(v) For all $k \in [J_-, J_+] \cap \mathbb{Z}$ we put $x_k = t_k$.

We use the symbol $CS(h, \rho, I)$ to denote the set of all ρ-covering sequences of $h \in Q_\rho(I)$. Note that the rule which was used to assign to a ρ-discretizing sequence of $h \in Q_\rho(I)$ the corresponding ρ-covering sequence defines in fact a one to one mapping between the sets $DS(h, \rho, I)$ and $CS(h, \rho, I)$. If $\alpha \in (1, +\infty)$, then we denote by $CS(h, \rho, I, \alpha)$ the subset of $CS(h, \rho, I)$ consisting of all ρ-covering sequences of $h \in Q_\rho(I)$ which correspond to elements of $DS(h, \rho, I, \alpha)$. To clarify our terminology, note the following. If $\{x_k\}_{k=K_-}^{K_+} \in CS(h, \rho, I)$, then the system of intervals $\{[x_{k-1}, x_k)\}_{k=K_-}^{K_+}$ forms a covering of the interval I.

In the following lemma we employ Convention 1.1.1 (iv).

Lemma 3.3.1 *Let $I = (a, b) \subseteq \mathbb{R}$, $h \in Q_\rho(I)$ and let $\{x_k\}_{k=K_-}^{K_+} \in CS(h, \rho, I)$. Put $\mathcal{Z} = \{k \in \mathbb{Z};\ K_- < k \leq K_+\}$. Then there is a decomposition*

$$\mathcal{Z} = \mathcal{Z}_1 \cup \mathcal{Z}_2, \quad \mathcal{Z}_1 \cap \mathcal{Z}_2 = \emptyset,$$

such that

$$h(x) \approx h(x_k) \quad \text{for all } x \in [x_{k-1}, x_k] \text{ and every } k \in \mathcal{Z}_1, \tag{3.3.1}$$

and

$$\left(\frac{\rho}{h}\right)(x) \approx \left(\frac{\rho}{h}\right)(x_k) \quad \text{for all } x \in [x_{k-1}, x_k] \text{ and every } k \in \mathcal{Z}_2. \tag{3.3.2}$$

Proof. Let $\{t_j\}_{j=J_-}^{J_+}$ be the ρ-discretizing sequence of h corresponding to the ρ-covering sequence $\{x_k\}_{k=K_-}^{K_+}$. Let $\mathcal{S}$, $\mathcal{S}_1$ and $\mathcal{S}_2$ be the sets from Lemma 3.2.5.

If $\mathcal{Z} = \mathcal{S}$, we put $\mathcal{Z}_1 = \mathcal{S}_1$, $\mathcal{Z}_2 = \mathcal{S}_2$ and both (3.3.1) and (3.3.2) hold due to (3.2.21) and (3.2.22).

Suppose now that $\mathcal{Z} \neq \mathcal{S}$. Then $K_+ > J_+$ or $K_- < J_-$.

If $K_+ > J_+$, then, by Remark 3.2.2, estimate (3.2.10) or (3.2.11) holds, which means that

$$h(t) \approx h(t_{J_+}) \quad \text{for all } t \in [t_{J_+}, b) \tag{3.3.3}$$

or

$$\left(\frac{\rho}{h}\right)(t) \approx \left(\frac{\rho}{h}\right)(t_{J_+}) \quad \text{for all } t \in [t_{J_+}, b). \tag{3.3.4}$$

If $K_- < J_-$, then by Remark 3.2.2, estimate (3.2.12) or (3.2.13) is satisfied, which means that

$$h(t) \approx h(t_{J_-}) \quad \text{for all } t \in (a, t_{J_-}] \tag{3.3.5}$$

or

$$\left(\frac{\rho}{h}\right)(t) \approx \left(\frac{\rho}{h}\right)(t_{J_-}) \quad \text{for all } t \in (a, t_{J_-}]. \tag{3.3.6}$$

Now we put
$\mathcal{Z}_1 = \mathcal{S}_1 \cup \{J_-, K_+\}$ if (3.3.3) and (3.3.5) hold;
$\mathcal{Z}_1 = \mathcal{S}_1 \cup \{J_-\}$ if (3.3.5) holds but (3.3.3) does not;
$\mathcal{Z}_1 = \mathcal{S}_1 \cup \{K_+\}$ if (3.3.3) holds but (3.3.5) does not.
Then, on taking $\mathcal{Z}_2 = \mathcal{Z} \setminus \mathcal{Z}_1$, we arrive at the desired result. □

Chapter 4

Discretization of weighted quasi-norms

4.1 Quasi-norms of ρ-quasiconcave functions

We start with some notation. Assume that $I = (a, b) \subseteq \mathbb{R}$, $\rho \in Ads(I)$, $\varphi \in Q_\rho(I)$ and $\{x_k\}_{k=K_-}^{K_+} \in CS(\varphi, \rho, I)$. Then we put

$$\begin{aligned}
\mathcal{K}_-^+ &= \{k \in \mathbb{Z};\ K_- \leq k \leq K_+\},\\
\mathcal{K}^+ &= \{k \in \mathbb{Z};\ K_- < k \leq K_+\},\\
\mathcal{K}_- &= \{k \in \mathbb{Z};\ K_- \leq k < K_+\},\\
\mathcal{K} &= \{k \in \mathbb{Z};\ K_- < k < K_+\}.
\end{aligned}$$

Lemma 4.1.1 *Let $I = (a, b) \subseteq \mathbb{R}$, $\rho \in Ads(I)$, $p \in (0, +\infty]$, $\mu \in \mathcal{B}^+(I)$ and $w \in \mathcal{W}(I, \mu)$. Put*

$$\varphi(x) := \|\min\{\rho(\cdot), \rho(x)\}\|_{p,w,I,\mu}, \quad x \in I, \tag{4.1.1}$$

and assume that

$$\varphi(\bar{x}) < +\infty \quad \text{for some} \quad \bar{x} \in I. \tag{4.1.2}$$

Let $\{x_k\}_{k=K_-}^{K_+} \in CS(\varphi, \rho, I, \alpha)$ with $\alpha > 2^{1/p}$. Then, for all $f \in Q_\rho(I)$,

$$\|f\|_{p,w,I,\mu} \approx \left\|\left(\frac{f\varphi}{\rho}\right)(x_k)\right\|_{\ell^p(\mathcal{K}_-^+)}, \tag{4.1.3}$$

where

$$\left(\frac{f\varphi}{\rho}\right)(x_k) := \lim_{x \to a+} \left(\frac{f\varphi}{\rho}\right)(x) \quad \text{if} \quad x_k = a \tag{4.1.4}$$

and

$$\left(\frac{f\varphi}{\rho}\right)(x_k) := \lim_{x \to b-} \left(\frac{f\varphi}{\rho}\right)(x) \quad \text{if} \quad x_k = b. \tag{4.1.5}$$

Proof. Let $f \in Q_\rho(I)$. First we prove that the limits in (4.1.4) and (4.1.5) exist. We have $x_k = a$ for some $k \in \mathcal{K}_-^+$ only if $K_- > -\infty$ (and then $x_{K_-} = a$). In such a case $J_- > -\infty$ as well, and, by Lemma 3.2.4 (v),

$$\lim_{x\to a+} \varphi(x) > 0 \qquad \text{or} \qquad \lim_{x\to a+} \left(\frac{\rho}{\varphi}\right)(x) > 0,$$

that is,

$$\lim_{x\to a+} \varphi(x) > 0 \qquad \text{or} \qquad \lim_{x\to a+} \left(\frac{\varphi}{\rho}\right)(x) < +\infty. \tag{4.1.6}$$

If $\lim_{x\to a+} \varphi(x) > 0$, then the limit (4.1.4) exist and

$$\lim_{x\to a+} \left(\frac{f\varphi}{\rho}\right)(x) = \lim_{x\to a+} \left(\frac{f}{\rho}\right)(x) \cdot \lim_{x\to a+} \varphi(x).$$

Suppose now that $\lim_{x\to a+} (\frac{\varphi}{\rho})(x) < +\infty$. Then the limit in (4.1.4) again exists and

$$\lim_{x\to a+} \left(\frac{f\varphi}{\rho}\right)(x) = \lim_{x\to a+} f(x) \cdot \lim_{x\to a+} \left(\frac{\varphi}{\rho}\right)(x).$$

The existence of the limits in (4.1.5) can be proved analogously.

To verify (4.1.3), we apply Lemma 3.3.1 (with $h = \varphi$). There is a decomposition

$$\mathcal{K}^+ = \mathcal{Z}_1 \cup \mathcal{Z}_2, \quad \mathcal{Z}_1 \cap \mathcal{Z}_2 = \emptyset,$$

such that

$$\varphi(x) \approx \varphi(x_k) \qquad \text{for all } x \in [x_{k-1}, x_k] \text{ and every } k \in \mathcal{Z}_1, \tag{4.1.7}$$

$$\left(\frac{\varphi}{\rho}\right)(x) \approx \left(\frac{\varphi}{\rho}\right)(x_k) \quad \text{for all } x \in [x_{k-1}, x_k] \text{ and every } k \in \mathcal{Z}_2.^1 \tag{4.1.8}$$

We shall distinguish two cases.

(i) Let $p \in (0, +\infty)$. Using the facts that

$$f \in \mathcal{M}^+(I, \mu; \uparrow) \qquad \text{and} \qquad \frac{f}{\rho} \in \mathcal{M}^+(I, \mu; \downarrow), \tag{4.1.9}$$

we obtain

$$\begin{aligned}
&\int_I f^p(t) w^p(t)\, d\mu \\
&= \sum_{k\in\mathcal{Z}_1} \int_{[x_{k-1},x_k)\cap I} f^p(t) w^p(t)\, d\mu + \sum_{k\in\mathcal{Z}_2} \int_{[x_{k-1},x_k)\cap I} f^p(t) w^p(t)\, d\mu \\
&\le \sum_{k\in\mathcal{Z}_1} \left(\frac{f}{\rho}\right)^p (x_{k-1}) \int_{[x_{k-1},x_k)\cap I} \rho^p(t) w^p(t)\, d\mu \\
&\quad + \sum_{k\in\mathcal{Z}_2} f^p(x_k) \int_{[x_{k-1},x_k)\cap I} w^p(t)\, d\mu.
\end{aligned}$$

[1] Recall that $\varphi(x_k) := \varphi(b-)$ and $(\frac{\varphi}{\rho})(x_k) := (\frac{\varphi}{\rho})(b-)$ if $x_k = b$. Similarly, $\varphi(x_{k-1}) := \varphi(a+)$ and $(\frac{\varphi}{\rho})(x_{k-1}) := (\frac{\varphi}{\rho})(a+)$ if $x_{k-1} = a$.

Since, by (4.1.1),

$$\int_{[x_{k-1},x_k)\cap I} \rho^p(t)w^p(t)\,d\mu \le \int_{(a,x_k)} \rho^p(t)w^p(t)\,d\mu \le \varphi^p(x_k)$$

and

$$\int_{[x_{k-1},x_k)\cap I} w^p(t)\,d\mu \le \int_{[x_{k-1},b)\cap I} w^p(t)\,d\mu \le \Big(\frac{\varphi}{\rho}\Big)^p(x_{k-1}),$$

we have

$$\int_I f^p(t)w^p(t)\,d\mu \le \sum_{k\in\mathcal{Z}_1}\Big(\frac{f}{\rho}\Big)^p(x_{k-1})\varphi^p(x_k) + \sum_{k\in\mathcal{Z}_2} f^p(x_k)\Big(\frac{\varphi}{\rho}\Big)^p(x_{k-1}).$$

On applying (4.1.7) and (4.1.8), we arrive at

$$\begin{aligned}\int_I f^p(t)w^p(t)\,d\mu &\lesssim \sum_{k\in\mathcal{Z}_1}\Big(\frac{f}{\rho}\Big)^p(x_{k-1})\varphi^p(x_{k-1}) + \sum_{k\in\mathcal{Z}_2} f^p(x_k)\Big(\frac{\varphi}{\rho}\Big)^p(x_k)\\ &\le 2\sum_{k\in\mathcal{K}_-^+}\Big(\frac{f\varphi}{\rho}\Big)^p(x_k),\end{aligned}$$

which implies that LHS(4.1.3) $\lesssim$ RHS(4.1.3).

To prove the reverse inequality, we write

$$\int_I f^p(t)w^p(t)\,d\mu = \sum_{k\in\mathcal{K}^+}\int_{[x_{k-1},x_k)\cap I} f^p(t)w^p(t)\,d\mu$$

and hence,

$$\begin{aligned}2\int_I f^p(t)w^p(t)\,d\mu &= \sum_{k\in\mathcal{K}^+}\int_{[x_{k-1},x_k)\cap I} f^p(t)w^p(t)\,d\mu\\ &\quad+\sum_{k\in\mathcal{K}_-}\int_{[x_k,x_{k+1})\cap I} f^p(t)w^p(t)\,d\mu.\end{aligned}$$

Therefore, using also (4.1.9) and the monotonicity of the function ρ, we obtain

$$\begin{aligned}2\int_I f^p(t)w^p(t)\,d\mu &\ge \sum_{k\in\mathcal{K}}\Big(\frac{f}{\rho}\Big)^p(x_k)\Big[\int_{[x_{k-1},x_k)\cap I}\rho^p(t)w^p(t)\,d\mu\\ &\quad+\rho^p(x_k)\int_{[x_k,x_{k+1})\cap I} w^p(t)d\mu\Big].\end{aligned}\tag{4.1.10}$$

Moreover, by (4.1.1),

$$\begin{aligned}&\int_{[x_{k-1},x_k)\cap I}\rho^p(t)w^p(t)\,d\mu + \rho^p(x_k)\int_{[x_k,x_{k+1})\cap I} w^p(t)\,d\mu\\ &\quad=\varphi^p(x_k) - \int_{(a,x_{k-1})}\rho^p(t)w^p(t)\,d\mu - \rho^p(x_k)\int_{[x_{k+1},b)} w^p(t)\,d\mu\\ &\quad\ge \varphi^p(x_k) - \varphi^p(x_{k-1}) - \rho^p(x_k)\frac{\varphi^p(x_{k+1})}{\rho^p(x_{k+1})}.\end{aligned}\tag{4.1.11}$$

By Definition 3.2.1,

$$\alpha^p \varphi^p(x_{k-1}) \le \varphi^p(x_k) \quad \text{if} \quad k \in \mathbb{Z},\ K_- + 1 < k \le K_+ - 1, \tag{4.1.12}$$

which yields that

$$\varphi^p(x_k) - \varphi^p(x_{k-1}) \ge \Big(1 - \frac{1}{\alpha^p}\Big)\varphi^p(x_k) \quad \text{if} \quad k \in \mathbb{Z},\ K_- + 1 < k \le K_+ - 1.$$

Similarly, the inequality

$$\alpha^p \Big(\frac{\rho}{\varphi}\Big)^p (x_k) \le \Big(\frac{\rho}{\varphi}\Big)^p (x_{k+1}) \quad \text{if} \quad k \in \mathbb{Z},\ K_- < k \le K_+ - 2,$$

implies that

$$\rho^p(x_k)\frac{\varphi^p(x_{k+1})}{\rho^p(x_{k+1})} \le \frac{1}{\alpha^p}\varphi^p(x_k) \quad \text{if} \quad k \in \mathbb{Z},\ K_- < k \le K_+ - 2. \tag{4.1.13}$$

Consequently,

$$\begin{aligned}\varphi^p(x_k) - \varphi^p(x_{k-1}) - \rho^p(x_k)&\frac{\varphi^p(x_{k+1})}{\rho^p(x_{k+1})} \\ &\ge \Big(1 - \frac{2}{\alpha^p}\Big)\varphi^p(x_k) \quad \text{if} \quad k \in \mathbb{Z},\ K_- + 2 \le k \le K_+ - 2.\end{aligned}$$

Together with (4.1.10) and (4.1.11), this shows that

$$\int_I f^p(t) w^p(t)\, d\mu \ge \frac{1}{2}\Big(1 - \frac{2}{\alpha^p}\Big) \sum_{K_- + 2 \le k \le K_+ - 2} \Big(\frac{f}{\rho}\Big)^p (x_k)\varphi^p(x_k) \tag{4.1.14}$$

and the inequality

$$\text{RHS(4.1.3)} \lesssim \text{LHS(4.1.3)} \tag{4.1.15}$$

follows if $K_- = -\infty$ and $K_+ = +\infty$.

To prove inequality (4.1.15) in the remaining cases, note that, for any $x \in I$,

$$\begin{aligned}\int_I f^p(t) w^p(t)\, d\mu &= \int_{(a,x)} f^p(t) w^p(t)\, d\mu + \int_{[x,b)} f^p(t) w^p(t)\, d\mu \\ &\ge \Big(\frac{f}{\rho}\Big)^p (x) \int_{(a,x)} \rho^p(t) w^p(t)\, d\mu + f^p(x)\int_{[x,b)} w^p(t)\, d\mu \\ &= \Big(\frac{f}{\rho}\Big)^p (x)\, \varphi^p(x) = \Big(\frac{f\varphi}{\rho}\Big)^p (x).\end{aligned} \tag{4.1.16}$$

Therefore, if $K_- > -\infty$ and $K_+ < +\infty$, then

$$\begin{aligned}&\int_I f^p(t) w^p(t)\, d\mu \\ &\ge \frac{1}{4}\Big[\Big(\frac{f\varphi}{\rho}\Big)^p (a+) + \Big(\frac{f\varphi}{\rho}\Big)^p (x_{K_- + 1}) + \Big(\frac{f\varphi}{\rho}\Big)^p (x_{K_+ - 1}) + \Big(\frac{f\varphi}{\rho}\Big)^p (b_+)\Big]\end{aligned} \tag{4.1.17}$$

and inequality (4.1.15) follows from (4.1.14) and (4.1.17). If $K_- = -\infty$ and $K_+ < +\infty$, or $K_- > -\infty$ and $K_+ = +\infty$, then the proof of (4.1.15) is analogous.

(ii) Let $p = +\infty$. Using (4.1.9), we obtain

$$\begin{aligned}\|f\|_{\infty,w,I,\mu} &= \sup_{k\in\mathcal{K}^+} \|f\|_{\infty,w,[x_{k-1},x_k)\cap I,\mu}\\ &= \max\Big\{\sup_{k\in\mathcal{Z}_1} \|f\|_{\infty,w,[x_{k-1},x_k)\cap I,\mu}, \sup_{k\in\mathcal{Z}_2} \|f\|_{\infty,w,[x_{k-1},x_k)\cap I,\mu}\Big\}\\ &\le \max\Big\{\sup_{k\in\mathcal{Z}_1} \Big(\frac{f}{\rho}\Big)(x_{k-1})\|\rho\|_{\infty,w,[x_{k-1},x_k)\cap I,\mu},\\ &\qquad \sup_{k\in\mathcal{Z}_2} f(x_k)\|1\|_{\infty,w,[x_{k-1},x_k)\cap I,\mu}\Big\}.\end{aligned}$$

Since, by (4.1.1),

$$\|\rho\|_{\infty,w,[x_{k-1},x_k)\cap I,\mu} \le \|\rho\|_{\infty,w,(a,x_k),\mu} \le \varphi(x_k)$$

and

$$\|1\|_{\infty,w,[x_{k-1},x_k),\mu} \le \|1\|_{\infty,w,[x_{k-1},b),\mu} \le \Big(\frac{\varphi}{\rho}\Big)(x_{k-1}),$$

we have

$$\|f\|_{\infty,w,I,\mu} \le \max\Big\{\sup_{k\in\mathcal{Z}_1} \Big(\frac{f}{\rho}\Big)(x_{k-1})\varphi(x_k), \ \sup_{k\in\mathcal{Z}_2} f(x_k)\Big(\frac{\varphi}{\rho}\Big)(x_{k-1})\Big\}.$$

On applying (4.1.7) and (4.1.8), we arrive at

$$\begin{aligned}\|f\|_{\infty,w,I,\mu} &\le \max\Big\{\sup_{k\in\mathcal{Z}_1} \Big(\frac{f}{\rho}\Big)(x_{k-1})\varphi(x_{k-1}), \ \sup_{k\in\mathcal{Z}_2} f(x_k)\Big(\frac{\varphi}{\rho}\Big)(x_k)\Big\}\\ &\le \sup_{k\in\mathcal{K}^+_-} \Big(\frac{f\varphi}{\rho}\Big)(x_k) = \Big\|\Big(\frac{f\varphi}{\rho}\Big)(x_k)\Big\|_{\ell^\infty(\mathcal{K}^+_-)},\end{aligned}$$

which implies that LHS(4.1.3) $\lesssim$ RHS(4.1.3).

To prove the reverse inequality, we write

$$\|f\|_{\infty,w,I,\mu} = \sup_{k\in\mathcal{K}^+} \|f\|_{\infty,w,[x_{k-1},x_k)\cap I,\mu} = \sup_{k\in\mathcal{K}_-} \|f\|_{\infty,w,[x_k,x_{k+1})\cap I,\mu},$$

and so

$$\|f\|_{\infty,w,I,\mu} = \sup_{k\in\mathcal{K}} \max\{\|f\|_{\infty,w,[x_{k-1},x_k)\cap I,\mu}, \ \|f\|_{\infty,w,[x_k,x_{k+1})\cap I,\mu}\}.$$

Therefore, using also (4.1.9), we obtain

$$\begin{aligned}&\|f\|_{\infty,w,I,\mu} \qquad (4.1.18)\\ &\ge \sup_{k\in\mathcal{K}} \max\Big\{\Big(\frac{f}{\rho}\Big)(x_k)\|\rho\|_{\infty,w,[x_{k-1},x_k)\cap I,\mu}, \ f(x_k)\|1\|_{\infty,w,[x_k,x_{k+1})\cap I,\mu}\Big\}\\ &= \sup_{k\in\mathcal{K}} \Big(\frac{f}{\rho}\Big)(x_k) \max\{\|\rho\|_{\infty,w,[x_{k-1},x_k)\cap I,\mu}, \rho(x_k)\|1\|_{\infty,w,[x_k,x_{k+1})\cap I,\mu}\}.\end{aligned}$$

Moreover, the triangle inequality yields

$$A_k := \|\rho\|_{\infty,w,[x_{k-1},x_k)\cap I,\mu} \geq \|\rho\|_{\infty,w,(a,x_k),\mu} - \|\rho\|_{\infty,w,(a,x_{k-1}),\mu} =: A_{1k} - A_{2k} \tag{4.1.19}$$

and

$$B_k := \rho(x_k)\|1\|_{\infty,w,[x_k,x_{k+1})\cap I,\mu} \geq \rho(x_k)\|1\|_{\infty,w,[x_k,b),\mu} - \rho(x_k)\|1\|_{\infty,w,[x_{k+1},b),\mu} =: B_{1k} - B_{2k}. \tag{4.1.20}$$

That is,

$$A_k \geq A_{1k} - A_{2k} \quad \text{and} \quad B_k \geq B_{1k} - B_{2k}. \tag{4.1.21}$$

Since, for all $k \in \mathcal{K}$,

$$A_{1k} - A_{2k} \geq A_{1k} - \max\{A_{2k}, B_{2k}\},$$

$$B_{1k} - B_{2k} \geq B_{1k} - \max\{A_{2k}, B_{2k}\},$$

we obtain

$$\max\{A_{1k}-A_{2k}, B_{1k}-B_{2k}\} \geq \max\{A_{1k}-\max\{A_{2k}, B_{2k}\}, B_{1k}-\max\{A_{2k}, B_{2k}\}\} = \max\{A_{1k}, B_{1k}\} - \max\{A_{2k}, B_{2k}\}.$$

Together with (4.1.21), this implies that, for all $k \in \mathcal{K}$,

$$\max\{A_k, B_k\} \geq \max\{A_{1k}, B_{1k}\} - \max\{A_{2k}, B_{2k}\}. \tag{4.1.22}$$

Moreover, by (4.1.1), for all $k \in \mathcal{K}$,

$$\max\{A_{1k}, B_{1k}\} = \varphi(x_k) \tag{4.1.23}$$

and

$$\max\{A_{2k}, B_{2k}\} \leq \max\left\{\varphi(x_{k-1}),\ \rho(x_k)\frac{\varphi(x_{k+1})}{\rho(x_{k+1})}\right\}.$$

Since, by Definition 3.2.1 (cf. (4.1.13)),

$$\rho(x_k)\frac{\varphi(x_{k+1})}{\rho(x_{k+1})} \leq \frac{1}{\alpha}\varphi(x_k) \quad \text{if} \quad k \in \mathbb{Z},\ K_- < k \leq K_+ - 2,$$

and (cf. (4.1.12))

$$\varphi(x_{k-1}) \leq \frac{1}{\alpha}\varphi(x_k) \quad \text{if} \quad k \in \mathbb{Z},\ K_- + 1 < k \leq K_+ - 1,$$

we have

$$\max\{A_{2k}, B_{2k}\} \leq \frac{1}{\alpha}\varphi(x_k) \quad \text{if} \quad k \in \mathbb{Z},\ K_- + 2 \leq k \leq K_+ - 2.$$

Together with (4.1.22) and (4.1.23), this yields

$$\text{(4.1.24)}\quad \max\{A_k, B_k\} \geq \Big(1 - \frac{1}{\alpha}\Big)\varphi(x_k) \quad \text{if} \quad k \in \mathbb{Z},\ K_- + 2 \leq k \leq K_+ - 2.$$

Thus, using (4.1.18), the definition of A_k and B_k, and (4.1.24), we arrive at

$$\text{(4.1.25)}\qquad \begin{aligned} \|f\|_{\infty,w,I,\mu} &\geq \sup_{K_-+2\leq k\leq K_+-2} \Big(\frac{f}{\rho}\Big)(x_k)\varphi(x_k)\Big(1 - \frac{1}{\alpha}\Big) \\ &\approx \sup_{K_-+2\leq k\leq K_+-2} \Big(\frac{f\varphi}{\rho}\Big)(x_k) \end{aligned}$$

and the inequality (4.1.15) follows if $K_- = -\infty$ and $K_+ = +\infty$.

To prove inequality (4.1.15) in the remaining cases, note that, for any $x \in I$,

$$\text{(4.1.26)}\qquad \begin{aligned} \|f\|_{\infty,w,I,\mu} &= \max\{\|f\|_{\infty,w,(a,x),\mu},\ \|f\|_{\infty,w,[x,b),\mu}\} \\ &\geq \max\Big\{\Big(\frac{f}{\rho}\Big)(x)\|\rho\|_{\infty,(a,x),\mu},\ f(x)\|1\|_{\infty,w,[x,b),\mu}\Big\} \\ &= \Big(\frac{f}{\rho}\Big)(x)\max\{\|\rho\|_{\infty,w,(a,x),\mu},\ \rho(x)\|1\|_{\infty,w,[x,b),\mu}\} \\ &= \Big(\frac{f}{\rho}\Big)(x)\varphi(x). \end{aligned}$$

Consequently, if $K_- > -\infty$ and $K_+ < +\infty$, then

$$\text{(4.1.27)}\quad \begin{aligned} &\|f\|_{\infty,w,I,\mu} \\ &\geq \max\Big\{\Big(\frac{f\varphi}{\rho}\Big)(a+), \Big(\frac{f\varphi}{\rho}\Big)(x_{K_-+1}), \Big(\frac{f\varphi}{\rho}\Big)(x_{K_+-1}), \Big(\frac{f\varphi}{\rho}\Big)(b-)\Big\} \end{aligned}$$

and the estimate (4.1.15) follows from (4.1.25) and (4.1.27). If $K_- = -\infty$ and $K_+ < +\infty$, or $K_- > -\infty$ and $K_+ = +\infty$, then the proof of (4.1.15) is analogous.

□

Remark 4.1.2 If assumption (4.1.2) is not satisfied, then

$$\text{(4.1.28)}\qquad \|f\|_{p,w,I,\mu} = +\infty \quad \text{for any} \quad f \in Q_\rho(I),\ f \not\equiv 0.$$

Indeed, we have

$$\text{(4.1.29)}\qquad \varphi = +\infty \quad \text{on} \quad I$$

and, since $f \in Q_\rho(I)$, $f \not\equiv 0$,

$$\text{(4.1.30)}\qquad f(x) > 0 \quad \text{for all } x \in I.$$

Take any $x \in I$. Then (4.1.16) or (4.1.26), together with (4.1.29) and (4.1.30), imply (4.1.28).

Convention 4.1.3 Let $I = (a,b) \subseteq \mathbb{R}$, $\rho \in Ads(I)$, $p \in (0,+\infty]$, $\mu \in \mathcal{B}^+(I)$, $w \in \mathcal{W}(I,\mu)$ and $f \in Q_\rho(I)$. Let φ be the ρ-fundamental function of the space $L^p(w,I,\mu)$ which satisfies (4.1.2) and let $\{x_k\}_{k=K_-}^{K_+} \in CS(\varphi,\rho,I)$. Then (in accordance with (4.1.4) and (4.1.5)) we put

$$\left(\frac{f\varphi}{\rho}\right)(x_k) := \lim_{x\to a+}\left(\frac{f\varphi}{\rho}\right)(x) =: A \quad \text{if } x_k = a \tag{4.1.31}$$

and

$$\left(\frac{f\varphi}{\rho}\right)(x_k) := \lim_{x\to b-}\left(\frac{f\varphi}{\rho}\right)(x) =: B \quad \text{if } x_k = b. \tag{4.1.32}$$

4.2 Quasi-norms of ρ-quasiconcave operators

Lemma 4.1.1 will now be applied to various ρ-quasiconcave operators.

Theorem 4.2.1 *Let $I = (a,b) \subseteq \mathbb{R}$, $\rho \in Ads(I)$, $q \in (0,+\infty]$, $\mu \in \mathcal{B}^+(I)$ and $w \in \mathcal{W}(I,\mu)$. Put*

$$\varphi(x) := \|\min\{\rho(\cdot),\rho(x)\}\|_{q,w,I,\mu},\ x \in I,$$

and suppose that

$$0 < \varphi(\bar{x}) < +\infty \quad \text{for some} \quad \bar{x} \in I. \tag{4.2.1}$$

Let $\{x_k\}_{k=K_-}^{K_+} \in CS(\varphi,\rho,I,\alpha)$ with $\alpha > 2^{1/q}$. Then, for all $g \in \mathcal{M}^+(I)$,

$$\begin{aligned}\left\| \sup_{a<t\le x} \rho(t) \int_t^b g(s)\,ds \right\|_{q,w,I,\mu} &\approx \left\| \frac{\varphi(x_k)}{\rho(x_k)} \sup_{a<t\le x_k} \rho(t) \int_t^b g(s)\,ds \right\|_{\ell^q(\mathcal{K}^\pm)} \\ &\approx \left\| \sup_{x_{k-1}<t\le x_k} \varphi(t) \int_t^{x_k} g(s)\,ds \right\|_{\ell^q(\mathcal{K}+)}.\end{aligned} \tag{4.2.2}$$

Remarks 4.2.2 (i) Assumption (4.2.1) implies that $0 < \varphi(x) < +\infty$ for all $x \in I$ (cf. (2.1.4) and (2.1.5) of Remarks 2.1.2 (i)).

If $\{x_k\}_{k=K_-}^{K_+} \in CS(\varphi,\rho,I)$ and $K_- > -\infty$, then (cf. (3.3.5) or (3.3.6) with φ instead of h)

$$\varphi(x) \approx \varphi(x_{J_-}) \quad \text{or} \quad \left(\frac{\rho}{\varphi}\right)(t) \approx \left(\frac{\rho}{\varphi}\right)(t_{J_-}) \qquad \text{for all } t \in (a, t_{J_-}].$$

Together with (4.2.1), this implies that $\varphi(x_{K_-}) > 0$.

Similarly one can prove that, if $\{x_k\}_{k=K_-}^{K_+} \in CS(\varphi,\rho,I)$, $K_+ < +\infty$ and (4.2.1) holds, then $\varphi(x_{K_+}) < +\infty$.

(ii) It follows from the proof of Theorem 4.2.1 given below that the second equivalence in (4.2.2) holds for any $\varphi \in Q_\rho(I)$, $0 \not\equiv \varphi \not\equiv +\infty$, such that $\{x_k\}_{k=K_-}^{K_+} \in CS(\varphi,\rho,I,\alpha)$ with $\alpha > 1$.

Proof of Theorem 4.2.1. Taking $g \in \mathcal{M}^+(I)$ and putting

$$f(x) := \sup_{a<t\leq x} \rho(t) \int_t^b g(s)\, ds, \quad x \in I, \tag{4.2.3}$$

we can see that the function f belongs to $Q_\rho(I)$ (cf. Corollary 2.2.10). Thus, the first equivalence in (4.2.2) is a consequence of Lemma 4.1.1. In other words,

$$\mathrm{LHS}(4.2.2) \approx A \sum_{k=K_-}^{K_+} \chi_{\{a\}}(x_k) + N + B \sum_{k=K_-}^{K_+} \chi_{\{b\}}(x_k), \tag{4.2.4}$$

where A and B are given by (4.1.31) and (4.1.32), respectively, and

$$N := \left\| \left(\frac{f\varphi}{\rho} \right)(x_k) \right\|_{\ell^q(\mathcal{K})}. \tag{4.2.5}$$

Now we are going to prove the second equivalence in (4.2.2). Since

$$(a, x_k] = \bigcup_{K_-<m\leq k} (x_{m-1}, x_m] \quad \text{for all } k \in \mathcal{K},$$

we have

$$f(x_k) = \sup_{a<t\leq x_k} \rho(t) \int_t^b g(s)\, ds = \sup_{K_-<m\leq k} a_m, \quad k \in \mathcal{K}, \tag{4.2.6}$$

where

$$a_m := \sup_{x_{m-1}<t\leq x_m} \rho(t) \int_t^b g(s)\, ds, \quad m \in \mathcal{K}. \tag{4.2.7}$$

Consequently,

$$N = \|\tau_k \sup_{K_-<m\leq k} a_m\|_{\ell^q(\mathcal{K})}, \tag{4.2.8}$$

where

$$\tau_k := \left(\frac{\varphi}{\rho} \right)(x_k), \quad k \in \mathcal{K}. \tag{4.2.9}$$

Since $\{x_k\}_{k=K_-}^{K_+} \in CS(\varphi, \rho, I)$, Definition 3.2.1 (i) implies that

$$\tau_k \leq \frac{1}{\alpha} \tau_{k-1} \quad \text{for all} \quad k \in \mathbb{Z},\ K_- + 2 \leq k \leq K_+ - 1, \tag{4.2.10}$$

which means that the sequence $\{\tau_k\}_{k\in\mathcal{K}}$ is geometrically decreasing. Therefore, by (1.3.4) of Lemma 1.3.4 (with $L = K_- + 1$ and $M = K_+ - 1$),

$$\|\tau_k \sup_{K_-<m\leq k} a_m\|_{\ell^q(\mathcal{K})} \approx \|\tau_k a_k\|_{\ell^q(\mathcal{K})}. \tag{4.2.11}$$

Together with (4.2.8), (4.2.7) and (4.2.9), this yields

$$N \approx \left\| \left(\frac{\varphi}{\rho}\right)(x_k) \sup_{x_{k-1}<t\le x_k} \rho(t) \int_t^b g(s)\,ds \right\|_{\ell^q(\mathcal{K})}. \tag{4.2.12}$$

Using the equality

$$\int_t^b g(s)\,ds = \int_t^{x_k} g(s)\,ds + \int_{x_k}^b g(s)\,ds \quad \text{for all } t \in (x_{k-1}, x_k] \tag{4.2.13}$$

and the fact that

$$\rho \in \mathcal{M}^+(I;\uparrow), \tag{4.2.14}$$

we obtain

$$N \approx \left\| \left(\frac{\varphi}{\rho}\right)(x_k) \sup_{x_{k-1}<t\le x_k} \rho(t) \int_t^{x_k} g(s)\,ds \right\|_{\ell^q(\mathcal{K})} + \left\| \varphi(x_k) \int_{x_k}^b g(s)\,ds \right\|_{\ell^q(\mathcal{K})}. \tag{4.2.15}$$

Put

$$b_m := \int_{x_m}^{x_{m+1}} g(s)\,ds \quad \text{and} \quad \sigma_m := \varphi(x_m), \quad m \in \mathcal{K}. \tag{4.2.16}$$

Then

$$\int_{x_k}^b g(s)\,ds = \sum_{m=k}^{K_+-1} b_m, \quad k \in \mathcal{K}. \tag{4.2.17}$$

Since (cf. Definition 3.2.1 (i))

$$\alpha\sigma_{k-1} \le \sigma_k \quad \text{for all} \quad k \in \mathbb{Z},\ K_- + 2 \le k \le K_+ - 1, \tag{4.2.18}$$

the sequence $\{\sigma_k\}_{k\in\mathcal{K}}$ is geometrically increasing. Thus, by (4.2.17), (4.2.16) and (1.3.5) of Lemma 1.3.5 (with $L = K_- + 1$ and $M = K_+ - 1$),

$$\begin{aligned} \left\| \varphi(x_k) \int_{x_k}^b g(s)\,ds \right\|_{\ell^q(\mathcal{K})} &\approx \left\| \sigma_k \sum_{m=k}^{K_+-1} b_m \right\|_{\ell^q(\mathcal{K})} \approx \|\sigma_k b_k\|_{\ell^q(\mathcal{K})} \\ &= \left\| \varphi(x_k) \int_{x_k}^{x_{k+1}} g(s)\,ds \right\|_{\ell^q(\mathcal{K})} \\ &= \left\| \varphi(x_{k-1}) \int_{x_{k-1}}^{x_k} g(s)\,ds \right\|_{\ell^q(\{\mathcal{K}_-+2,\ldots,K_+\})}. \end{aligned} \tag{4.2.19}$$

Together with (4.2.15), this implies that

$$\begin{aligned} N \approx{}& \left\| \left(\frac{\varphi}{\rho}\right)(x_k) \sup_{x_{k-1}<t\le x_k} \rho(t) \int_t^{x_k} g(s)\,ds \right\|_{\ell^q(\mathcal{K})} \\ &+ \left\| \varphi(x_{k-1}) \int_{x_{k-1}}^{x_k} g(s)\,ds \right\|_{\ell^q(\{K_-+2,\ldots,K_+\})}. \end{aligned} \tag{4.2.20}$$

Since $\varphi \in Q_\rho(I)$, we can apply Lemma 3.3.1 to get a decomposition of the set $\mathcal{K}^+$,

$$\mathcal{K}^+ = \mathcal{Z}_1 \cup \mathcal{Z}_2, \quad \mathcal{Z}_1 \cap \mathcal{Z}_2 = \emptyset, \tag{4.2.21}$$

such that (4.1.7) and (4.1.8) hold. Using (4.2.14) and (4.1.7), we observe that

$$\begin{aligned} &\left\| \left(\frac{\varphi}{\rho}\right)(x_k) \sup_{x_{k-1}<t\le x_k} \rho(t) \int_t^{x_k} g(s)\,ds \right\|_{\ell^q(\mathcal{Z}_1)} \\ &\quad \le \left\| \varphi(x_k) \int_{x_{k-1}}^{x_k} g(s)\,ds \right\|_{\ell^q(\mathcal{Z}_1)} \approx \left\| \varphi(x_{k-1}) \int_{x_{k-1}}^{x_k} g(s)\,ds \right\|_{\ell^q(\mathcal{Z}_1)}. \end{aligned} \tag{4.2.22}$$

By (4.1.8), and since $\rho \in C(I)$, we obtain

$$\begin{aligned} &\left\| \varphi(x_{k-1}) \int_{x_{k-1}}^{x_k} g(s)\,ds \right\|_{\ell^q(\mathcal{Z}_2)} \\ &\quad = \left\| \left(\frac{\varphi}{\rho}\right)(x_{k-1})\rho(x_{k-1}) \int_{x_{k-1}}^{x_k} g(s)\,ds \right\|_{\ell^q(\mathcal{Z}_2)} \\ &\quad \lesssim \left\| \left(\frac{\varphi}{\rho}\right)(x_k) \sup_{x_{k-1}<t\le x_k} \rho(t) \int_t^{x_k} g(s)\,ds \right\|_{\ell^q(\mathcal{Z}_2)}. \end{aligned} \tag{4.2.23}$$

Moreover, by (4.1.8),

$$\begin{aligned} &\left(\frac{\varphi}{\rho}\right)(x_k) \sup_{x_{k-1}<t\le x_k} \rho(t) \int_t^{x_k} g(s)\,ds \\ &\quad \approx \sup_{x_{k-1}<t\le x_k} \varphi(t) \int_t^{x_k} g(s)\,ds \quad \text{for all } k \in \mathcal{Z}_2 \end{aligned} \tag{4.2.24}$$

and, by (4.1.7),

$$\varphi(x_{k-1}) \int_{x_{k-1}}^{x_k} g(s)\,ds \approx \sup_{x_{k-1}<t\le x_k} \varphi(t) \int_t^{x_k} g(s)\,ds \quad \text{for all } k \in \mathcal{Z}_1. \tag{4.2.25}$$

(i) Assume first that

$$K_- = -\infty \quad \text{and} \quad K_+ = +\infty. \tag{4.2.26}$$

Then $\mathcal{K} = \mathcal{K}^+ = \mathbb{Z}$ and, by (4.2.20),

$$\begin{aligned} N \approx &\left\| \left(\frac{\varphi}{\rho}\right)(x_k) \sup_{x_{k-1}<t\le x_k} \rho(t) \int_t^{x_k} g(s)\,ds \right\|_{\ell^q(\mathcal{K}^+)} \\ &+ \left\| \varphi(x_{k-1}) \int_{x_{k-1}}^{x_k} g(s)\,ds \right\|_{\ell^q(\mathcal{K}^+)}. \end{aligned} \tag{4.2.27}$$

Together with (4.2.21), estimates (4.2.27), (4.2.22) and (4.2.23) imply that

$$\begin{aligned} N \approx &\left\| \left(\frac{\varphi}{\rho}\right)(x_k) \sup_{x_{k-1}<t\le x_k} \rho(t) \int_t^{x_k} g(s)\,ds \right\|_{\ell^q(\mathcal{Z}_2)} \\ &+ \left\| \varphi(x_{k-1}) \int_{x_{k-1}}^{x_k} g(s)\,ds \right\|_{\ell^q(\mathcal{Z}_1)}. \end{aligned} \tag{4.2.28}$$

Combining estimates (4.2.28), (4.2.24), (4.2.25) and using (4.2.21), we arrive at

$$N \approx \Big\| \sup_{x_{k-1}<t\le x_k} \varphi(t) \int_t^{x_k} g(s)\,ds \Big\|_{\ell^q(\mathcal{K}^+)}. \tag{4.2.29}$$

Since assumption (4.2.26) implies that RHS(4.2.4) $= N$, the second estimate in (4.2.2) follows from (4.2.29) and (4.2.4).

(ii) Assume now that

$$K_- > -\infty \quad \text{and} \quad K_+ = +\infty. \tag{4.2.30}$$

Then

$$\mathcal{K} = \mathcal{K}^+ \tag{4.2.31}$$

and

$$\text{RHS}(4.2.4) = A + N. \tag{4.2.32}$$

Using (4.2.3), (4.1.31) and

$$\frac{\varphi}{\rho} \in \mathcal{M}^+(I;\downarrow), \tag{4.2.33}$$

we see that

$$\begin{aligned} A &= \lim_{x\to a+} \frac{\varphi(x)}{\rho(x)} \sup_{a<t\le x} \rho(t) \int_t^b g(s)\,ds \\ &\le \lim_{x\to a+} \sup_{a<t\le x} \varphi(t) \int_t^b g(s)\,ds \\ &\le \sup_{a<t\le x_{K_-+1}} \varphi(t) \int_t^b g(s)\,ds. \end{aligned}$$

Moreover, applying (4.2.13) with $k = K_- + 1$ and the fact that

$$\varphi \in \mathcal{M}^+(I;\uparrow), \tag{4.2.34}$$

we obtain

$$A \le \sup_{a<t\le x_{K_-+1}} \varphi(t) \int_t^{x_{K_-+1}} g(s)\,ds + \varphi(x_{K_-+1}) \int_{x_{K_-+1}}^b g(s)\,ds =: A_1 + A_2.$$

Since, by (4.2.19) and (4.2.20), $A_2 \lesssim N$, we arrive at

$$A \lesssim A_1 + N. \tag{4.2.35}$$

On the other hand, using (4.2.14) and (4.2.34), we have

$$
\begin{aligned}
A &\geq \lim_{x\to a+} \frac{\varphi(x)}{\rho(x)}\Big(\int_x^{x_{K_-+1}} g(s)\,ds\Big) \sup_{a<t\leq x} \rho(t) \\
&= \lim_{x\to a+} \varphi(x) \int_x^{x_{K_-+1}} g(s)\,ds \\
&\geq \varphi(x_{K_-}) \lim_{x\to a+} \int_x^{x_{K_-+1}} g(s)\,ds \\
&= \varphi(x_{K_-}) \int_{x_{K_-}}^{x_{K_-+1}} g(s)\,ds \\
&=: A_0
\end{aligned} \tag{4.2.36}
$$

(note that the assumptions $K_- > -\infty$ and (4.2.1) imply that $\varphi(x_{K_-}) > 0$ - cf. Remarks 4.2.2 (i)). By (4.2.36), (4.2.32) and (4.2.35),

$$A_0 + N \leq A + N = \mathrm{RHS}(4.2.4) \lesssim A_1 + N. \tag{4.2.37}$$

Now (4.2.31), (4.2.21), (4.2.20), (4.2.22) and (4.2.23) show that

$$N \lesssim \mathrm{RHS}(4.2.28),$$

and, on using (4.2.24) and (4.2.25), we obtain that

$$N \lesssim \mathrm{RHS}(4.2.29). \tag{4.2.38}$$

Moreover,

$$A_1 \lesssim \mathrm{RHS}(4.2.29).$$

Together with (4.2.37) and (4.2.38), this implies that

$$\mathrm{RHS}(4.2.4) \lesssim \mathrm{RHS}(4.2.29). \tag{4.2.39}$$

On the other hand, we deduce from (4.2.20), the definition of A_0 (see (4.2.36)) and (4.2.30) that

$$
\begin{aligned}
A_0 + N \approx & \left\| \Big(\frac{\varphi}{\rho}\Big)(x_k) \sup_{x_{k-1}<t\leq x_k} \rho(t) \int_t^{x_k} g(s)\,ds \right\|_{\ell^q(\{K_-+1,K_-+2,\dots\})} \\
&+ \left\| \varphi(x_{k-1}) \int_{x_{k-1}}^{x_k} g(s)\,ds \right\|_{\ell^q(\{K_-+1,K_-+2,\dots\})}
\end{aligned}
$$

so that

$$A_0 + N \approx \mathrm{RHS}(4.2.27).$$

Consequently, applying (4.2.22), (4.2.23), (4.2.24) and (4.2.25), we arrive at

$$A_0 + N \approx \mathrm{RHS}(4.2.29). \tag{4.2.40}$$

Now, (4.2.39), (4.2.37) and (4.2.40) imply that RHS(4.2.4) $\approx$ RHS(4.2.29), and so the second estimate in (4.2.2) follows from (4.2.32) and the definition of RHS(4.2.29).

(iii) Assume now that

$$K_- = -\infty \quad \text{and} \quad K_+ < +\infty. \tag{4.2.41}$$

(Note that $\mathcal{K} \subsetneqq \mathcal{K}^+$ now.) Then

$$\text{RHS}(4.2.4) = N + B. \tag{4.2.42}$$

By (4.1.32), (4.2.3) and (4.2.33),

$$\begin{aligned} B &= \lim_{x \to b-} \frac{\varphi(x)}{\rho(x)} \sup_{a<t\le x} \rho(t) \int_t^b g(s)\,ds \\ &\le \lim_{x \to b-} \sup_{a<t\le x} \varphi(t) \int_t^b g(s)\,ds \\ &= \sup_{a<t\le b} \varphi(t) \int_t^b g(s)\,ds. \end{aligned} \tag{4.2.43}$$

Moreover,

$$(a,b] = \bigcup_{K_- < k \le K_+} (x_{k-1}, x_k]$$

and so

$$\begin{aligned} \text{RHS}(4.2.43) &= \sup_{K_- < k \le K_+} \sup_{x_{k-1} < t \le x_k} \varphi(t) \int_t^b g(s)\,ds \\ &= \max\Bigg\{ \sup_{x_{K_+-1} < t \le x_{K_+}} \varphi(t) \int_t^b g(s)\,ds, \\ &\qquad \sup_{K_- < k < K_+} \sup_{x_{k-1} < t \le x_k} \varphi(t) \int_t^b g(s)\,ds \Bigg\}. \end{aligned} \tag{4.2.44}$$

Applying (4.2.13) (with $k \in \mathcal{K}$, i.e. $K_- < k < K_+$) and (4.2.34), we see that

$$\begin{aligned} \text{RHS}(4.2.44) &\le \sup_{x_{K_+-1} < t \le x_{K_+}} \varphi(t) \int_t^b g(s)ds \\ &\quad + \sup_{k\in\mathcal{K}} \sup_{x_{k-1} < t \le x_k} \varphi(t) \int_t^{x_k} g(s)\,ds + \sup_{k\in\mathcal{K}} \varphi(x_k) \int_{x_k}^b g(s)\,ds \\ &=: B_1 + B_2 + B_3. \end{aligned} \tag{4.2.45}$$

In view of (4.2.41), $K_- + 1 = K_- + 2 = -\infty$ and so we have from (4.2.24), (4.2.25) and (4.2.20) that

$$B_2 \lesssim N.$$

Similarly, by (4.2.15),

$$B_3 \lesssim N.$$

Therefore,

$$B \lesssim B_1 + N. \tag{4.2.46}$$

Assume now that

$$K_+ \in \mathcal{Z}_1. \tag{4.2.47}$$

Then (4.2.25) with $k = K_+$ and (4.2.20) (note that $b = x_{K_+}$) imply that

$$B_1 \lesssim N. \tag{4.2.48}$$

Together with (4.2.46), this shows that $B \lesssim N$. The last estimate and (4.2.42) yield

$$\mathrm{RHS}(4.2.4) \approx N. \tag{4.2.49}$$

Using (4.2.41), (4.2.47) and (4.2.20), we obtain

$$N \gtrsim \left\| \left(\frac{\varphi}{\rho}\right)(x_k) \sup_{x_{k-1}<t\le x_k} \rho(t) \int_t^{x_k} g(s)\,ds \right\|_{\ell^q(\mathcal{Z}_2)} + \left\| \varphi(x_{k-1}) \int_{x_{k-1}}^{x_k} g(s)\,ds \right\|_{\ell^q(\mathcal{Z}_1)}.$$

By (4.2.20), (4.2.22) and (4.2.23), the reverse estimate holds as well. Consequently, we have arrived at (4.2.28), which, together with (4.2.24), (4.2.25) and (4.2.21), implies (4.2.29). The second estimate in (4.2.2) follows from (4.2.29), (4.2.49) and (4.2.4).

Assume now that

$$K_+ \in \mathcal{Z}_2. \tag{4.2.50}$$

Then, by (4.1.32), (4.2.3) and (4.1.8),

$$\begin{aligned} B &\ge \lim_{x\to b-} \frac{\varphi(x)}{\rho(x)} \sup_{x_{K_+-1}<t\le x} \rho(t) \int_t^b g(s)\,ds \\ &\approx \lim_{x\to b-} \sup_{x_{K_+-1}<t\le x} \varphi(t) \int_t^b g(s)\,ds \\ &= \sup_{x_{K_+-1}<t\le x_{K_+}} \varphi(t) \int_t^b g(s)\,ds \\ &= B_1. \end{aligned} \tag{4.2.51}$$

Using (4.2.42), (4.2.46) and (4.2.51), we arrive at

$$\mathrm{RHS}(4.2.4) \approx B_1 + N,$$

which, together with (4.2.20), (4.2.24) (for $k = K_+$), the equality $b = x_{K_+}$ and (4.2.41), shows that

$$\text{RHS}(4.2.4) \approx \left\| \left(\frac{\varphi}{\rho}\right)(x_k) \sup_{x_{k-1}<t\le x_k} \rho(t) \int_t^{x_k} g(s)\, ds \right\|_{\ell^q(\mathcal{K}^+)} + \left\| \varphi(x_{k-1}) \int_{x_{k-1}}^{x_k} g(s)\, ds \right\|_{\ell^q(\mathcal{K}^+)}$$

(recall that $\{K_- + 2, ..., K_+\} = \mathcal{K}^+$ since $K_- = -\infty$). Hence, just as in the case (i), we obtain the result.

(iv) Finally, suppose that $K_- > -\infty$ and $K_+ < +\infty$. The desired result is proved by arguments similar to those used in the other cases above. □

In Theorem 4.2.1 we have discretized a weighted quasi-norm of the function given by (2.2.42). Our next assertion provides such a result for the function defined by (2.2.43).

Theorem 4.2.3 *Let $I = (a, b) \subseteq \mathbb{R}$, $\rho \in Ads(I)$, $q \in (0, +\infty]$, $\mu \in \mathcal{B}^+(I)$ and let $w \in \mathcal{W}(I, \mu)$. Put*

$$\varphi(x) := \|\min\{\rho(\cdot), \rho(x)\}\|_{q,w,I,\mu}, \quad x \in I,$$

and suppose that

$$0 < \varphi(\bar{x}) < +\infty \text{ for some } \bar{x} \in I. \tag{4.2.52}$$

Let $\{x_k\}_{k=K_-}^{K_+} \in CS(\varphi, \rho, I, \alpha)$ with $\alpha > 2^{1/q}$. Then, for all $g \in \mathcal{M}^+(I)$,

$$\begin{aligned} &\| \|\rho(t)\|g\|_{\infty,(t,b)}\|_{\infty,(a,x)}\|_{q,w,I,\mu} \\ &\approx \left\| \frac{\varphi(x_k)}{\rho(x_k)} \|\rho(t)\|g\|_{\infty,(t,b)}\|_{\infty,(a,x_k)} \right\|_{\ell^q(\mathcal{K}_-^+)} \\ &\approx \| \|\varphi(t)\|g\|_{\infty,(t,x_k)}\|_{\infty,(x_{k-1},x_k)}\|_{\ell^q(\mathcal{K}^+)}. \end{aligned} \tag{4.2.53}$$

Proof. It is analogous to that of Theorem 4.2.1 and so left to the reader. □

Remarks 4.2.4 (i) Let all the assumptions of Theorem 4.2.3 be satisfied. By Remark 2.2.13(i),

$$\|\rho(t)\|g\|_{\infty,(t,b)}\|_{\infty,(a,x)} = \|\min\{\rho(\cdot), \rho(x)\}\|_{\infty,g,I} \quad \text{for all} \quad x \in I,$$

and, similarly, exchanging the order of essential suprema, we obtain

$$\|\varphi(t)\|g\|_{\infty,(t,x_k)}\|_{\infty,(x_{k-1},x_k)} = \|\varphi g\|_{\infty,(x_{k-1},x_k)} = \|\varphi\|_{\infty,g,(x_{k-1},x_k)},$$

for all $k \in \mathcal{K}^+$. Thus, (4.2.53) can be rewritten as

$$\begin{aligned} &\| \|\min\{\rho(\cdot), \rho(x)\}\|_{\infty,g,I}\|_{q,w,I,\mu} \\ &\approx \left\| \frac{\varphi(x_k)}{\rho(x_k)} \|\min\{\rho(\cdot), \rho(x_k)\}\|_{\infty,g,I} \right\|_{\ell^q(\mathcal{K}_-^+)} \\ &\approx \| \|\varphi g\|_{\infty,(x_{k-1},x_k)}\|_{\ell^q(\mathcal{K}^+)} \\ &= \| \|\varphi\|_{\infty,g,(x_{k-1},x_k)}\|_{\ell^q(\mathcal{K}^+)}. \end{aligned} \tag{4.2.54}$$

(ii) Assumption (4.2.52) implies that $0 < \varphi(x) < +\infty$ for all $x \in I$.

(iii) It follows from the proof of Theorem 4.2.3 that the second equivalence in (4.2.53) holds for any $\varphi \in Q_\rho(I)$, $0 \not\equiv \varphi \not\equiv +\infty$, such that $\{x_k\}_{k=K_-}^{K_+} \in CS(\varphi, \rho, I, \alpha)$ with $\alpha > 1$.

Theorem 4.2.5 *Let $I = (a, b) \subseteq \mathbb{R}$, $\rho \in Ads(I)$, $q \in (0, +\infty]$, $\mu \in \mathcal{B}^+(I)$ and let $w \in \mathcal{W}(I, \mu)$. Put*

$$\varphi(x) := \| \min\{\rho(\cdot), \rho(x)\}\|_{q,w,I,\mu}, \quad x \in I, \tag{4.2.55}$$

and assume that

$$0 < \varphi(\bar{x}) < +\infty \quad \text{for some} \quad \bar{x} \in I. \tag{4.2.56}$$

Let $\{x_k\}_{k=K_-}^{K_+} \in CS(\varphi, \rho, I, \alpha)$ with $\alpha > 2^{1/q}$. Then, for all $\nu \in \mathcal{B}^+(I)$,

$$\begin{aligned} &\| \, \| \min\{\rho(\cdot), \rho(x)\}\|_{1,I,\nu}\|_{q,w,I,\mu} \\ &\approx \left\| \frac{\varphi(x_k)}{\rho(x_k)} \, \| \min\{\rho(\cdot), \rho(x_k)\}\|_{1,I,\nu} \right\|_{\ell^q(\mathcal{K}_-^+)} \\ &\approx \left\| \int_{(x_{k-1}, x_k]} \varphi(t) \, d\bar{\nu} \right\|_{\ell^q(\mathcal{K}^+)}, \end{aligned} \tag{4.2.57}$$

where $\bar{\nu}$ is the extension of ν by zero in $\mathbb{R} \setminus I$ (cf. page 2).

Remark 4.2.6 It follows from the proof of Theorem 4.2.5 given below that the second equivalence in (4.2.57) holds if φ is replaced by any $\varphi \in Q_\rho$, $0 \not\equiv \varphi \not\equiv +\infty$, which is such that $\{x_k\}_{k=K_-}^{K_+} \in CS(\varphi, \rho, I, \alpha)$ with $\alpha > 1$.

Proof of Theorem 4.2.5. Put

$$f(x) := \| \min\{\rho(\cdot), \rho(x)\}\|_{1,I,\nu}, \quad x \in I. \tag{4.2.58}$$

The function f belongs to $Q_\rho(I)$ (cf. Example 2.2.1 (ii)). Thus, the first equivalence in (4.2.57) is a consequence of Lemma 4.1.1. In other words,

$$\text{LHS}(4.2.57) \approx A \sum_{k=K_-}^{K_+} \chi_{\{a\}}(x_k) + N + B \sum_{k=K_-}^{K_+} \chi_{\{b\}}(x_k), \tag{4.2.59}$$

where A and B are given by (4.1.31) and (4.1.32), respectively, and

$$N := \left\| \left(\frac{f\varphi}{\rho} \right)(x_k) \right\|_{\ell^q(\mathcal{K})}. \tag{4.2.60}$$

Since, by (4.2.58),

$$f(x_k) = \int_{(a, x_k]} \rho(t) \, d\nu + \rho(x_k) \int_{(x_k, b)} d\nu, \tag{4.2.61}$$

we have

$$(4.2.62)\quad N \approx \left\| \left(\frac{\varphi}{\rho}\right)(x_k) \int_{(a,x_k]} \rho(t)\,d\nu \right\|_{\ell^q(\mathcal{K})} + \left\| \varphi(x_k) \int_{(x_k,b)} d\nu \right\|_{\ell^q(\mathcal{K})} =: N_1 + N_2.$$

Moreover,

$$(4.2.63)\qquad \int_{(a,x_k]} \rho(t)\,d\nu = \sum_{K_-<m\le k} a_m, \quad k \in \mathcal{K},$$

where

$$(4.2.64)\qquad a_m = \int_{(x_{m-1},x_m]} \rho(t)\,d\nu, \quad m \in \mathcal{K}.$$

Consequently,

$$(4.2.65)\qquad N_1 = \left\| \tau_k \sum_{K_-<m\le k} a_m \right\|_{\ell^q(\mathcal{K})},$$

where the sequence $\{\tau_k\}_{k\in\mathcal{K}}$ is given by (4.2.9). By (4.2.10), $\{\tau_k\}_{k\in\mathcal{K}}$ is geometrically decreasing. Therefore, by (1.3.3) of Lemma 1.3.4 (with $L = K_- + 1$ and $M = K_+ - 1$),

$$(4.2.66)\qquad \left\| \tau_k \sum_{K_-<m\le k} a_m \right\|_{\ell^q(\mathcal{K})} \approx \|\tau_k a_k\|_{\ell^q(\mathcal{K})}.$$

Together with (4.2.65), (4.2.64) and (4.2.9), this yields

$$(4.2.67)\qquad N_1 \approx \left\| \left(\frac{\varphi}{\rho}\right)(x_k) \int_{(x_{k-1},x_k]} \rho(t)\,d\nu \right\|_{\ell^q(\mathcal{K})}.$$

Put

$$(4.2.68)\qquad b_m = \int_{(x_m,x_{m+1}]} d\bar{\nu} \quad \text{and} \quad \sigma_m = \varphi(x_m), \quad m \in \mathcal{K}.$$

Then

$$(4.2.69)\qquad \int_{(x_k,b)} d\nu = \sum_{m=k}^{K_+-1} b_m, \quad k \in \mathcal{K}.$$

Since (cf. (4.2.18)) the sequence $\{\sigma_k\}_{k\in\mathcal{K}}$ is geometrically increasing, we obtain from the definition of N_2, (4.2.69), (4.2.68) and (1.3.5) of Lemma 1.3.5 (with $L = K_- + 1$ and $M = K_+ - 1$),

$$(4.2.70)\qquad \begin{aligned} N_2 &= \left\| \varphi(x_k) \int_{(x_k,b)} d\nu \right\|_{\ell^q(\mathcal{K})} = \left\| \sigma_k \sum_{m=k}^{K_+-1} b_m \right\|_{\ell^q(\mathcal{K})} \\ &\approx \|\sigma_k b_k\|_{\ell^q(\mathcal{K})} = \|\varphi(x_k) \int_{(x_k,x_{k+1}]} d\bar{\nu}\|_{\ell^q(\mathcal{K})} \\ &= \left\| \varphi(x_{k-1}) \int_{(x_{k-1},x_k]} d\bar{\nu} \right\|_{\ell^q(\{K_-+2,\dots,K_+\})}. \end{aligned}$$

Together with (4.2.67) and (4.2.62), this implies that

$$N \approx \left\| \left(\frac{\varphi}{\rho}\right)(x_k) \int_{(x_{k-1},x_k]} \rho(t)\, d\bar{\nu} \right\|_{\ell^q(\mathcal{K})} + \left\| \varphi(x_{k-1}) \int_{(x_{k-1},x_k]} d\bar{\nu} \right\|_{\ell^q(\{K_-+2,\dots,K_+\})}. \tag{4.2.71}$$

Since $\varphi \in Q_\rho(I)$, we can apply Lemma 3.3.1 to get a decomposition of the set $\mathcal{K}^+$,

$$\mathcal{K}^+ = \mathcal{Z}_1 \cup \mathcal{Z}_2, \quad \mathcal{Z}_1 \cap \mathcal{Z}_2 = \emptyset, \tag{4.2.72}$$

such that (4.1.7) and (4.1.8) hold. Using (4.2.14) and (4.1.7), we observe that

$$\begin{aligned} &\left\| \left(\frac{\varphi}{\rho}\right)(x_k) \int_{(x_{k-1},x_k]} \rho(t)\, d\bar{\nu} \right\|_{\ell^q(\mathcal{Z}_1)} \\ &\leq \left\| \varphi(x_k) \int_{(x_{k-1},x_k]} d\bar{\nu} \right\|_{\ell^q(\mathcal{Z}_1)} \\ &\approx \left\| \varphi(x_{k-1}) \int_{(x_{k-1},x_k]} d\bar{\nu} \right\|_{\ell^q(\mathcal{Z}_1)}. \end{aligned} \tag{4.2.73}$$

Applying (4.2.14) and (4.1.8), we obtain

$$\begin{aligned} \left\| \varphi(x_{k-1}) \int_{(x_{k-1},x_k]} d\bar{\nu} \right\|_{\ell^q(\mathcal{Z}_2)} &\leq \left\| \left(\frac{\varphi}{\rho}\right)(x_{k-1}) \int_{(x_{k-1},x_k]} \rho\, d\bar{\nu} \right\|_{\ell^q(\mathcal{Z}_2)} \\ &\approx \left\| \left(\frac{\varphi}{\rho}\right)(x_k) \int_{(x_{k-1},x_k]} \rho\, d\bar{\nu} \right\|_{\ell^q(\mathcal{Z}_2)}. \end{aligned} \tag{4.2.74}$$

Moreover, by (4.1.8),

$$\left(\frac{\varphi}{\rho}\right)(x_k) \int_{(x_{k-1},x_k]} \rho(t)\, d\bar{\nu} \approx \int_{(x_{k-1},x_k]} \varphi(t)\, d\bar{\nu} \quad \text{for all } k \in \mathcal{Z}_2 \tag{4.2.75}$$

and, by (4.1.7),

$$\varphi(x_{k-1}) \int_{(x_{k-1},x_k]} d\bar{\nu} \approx \int_{(x_{k-1},x_k]} \varphi(t)\, d\bar{\nu} \quad \text{for all } k \in \mathcal{Z}_1. \tag{4.2.76}$$

(i) Assume first that

$$K_- = -\infty \quad \text{and} \quad K_+ = +\infty. \tag{4.2.77}$$

Then $\mathcal{K} = \mathcal{K}^+ = \mathcal{Z}$ and, by (4.2.71),

$$N \approx \left\| \left(\frac{\varphi}{\rho}\right)(x_k) \int_{(x_{k-1},x_k]} \rho(t)\, d\bar{\nu} \right\|_{\ell^q(\mathcal{K}^+)} + \left\| \varphi(x_{k-1}) \int_{(x_{k-1},x_k]} d\bar{\nu} \right\|_{\ell^q(\mathcal{K}^+)}. \tag{4.2.78}$$

Together with (4.2.72), estimates (4.2.78), (4.2.73) and (4.2.74) imply that

$$N \approx \left\| \left(\frac{\varphi}{\rho}\right)(x_k) \int_{(x_{k-1},x_k]} \rho(t)\, d\bar{\nu} \right\|_{\ell^q(\mathcal{Z}_2)} + \left\| \varphi(x_{k-1}) \int_{(x_{k-1},x_k]} d\bar{\nu} \right\|_{\ell^q(\mathcal{Z}_1)}. \tag{4.2.79}$$

Combining estimates (4.2.79), (4.2.75), (4.2.76), and using (4.2.72), we arrive at

$$N \approx \left\| \int_{(x_{k-1},x_k]} \varphi(t)\, d\bar{\nu} \right\|_{\ell^q \mathcal{K}^+)}. \tag{4.2.80}$$

Since assumption (4.2.77) implies that RHS(4.2.59) $= N$, the second estimate in (4.2.57) follows from (4.2.80) and (4.2.59).

(ii) Assume now that

$$K_- > -\infty \quad \text{and} \quad K_+ = +\infty. \tag{4.2.81}$$

Then

$$\mathcal{K} = \mathcal{K}^+ \tag{4.2.82}$$

and

$$\text{RHS(4.2.59)} = A + N. \tag{4.2.83}$$

Using (4.1.31), (4.2.58) and (4.2.33), we see that

$$\begin{aligned} A &= \lim_{x\to a+} \frac{\varphi(x)}{\rho(x)} \| \min\{\rho(\cdot), \rho(x)\}\|_{1,I,\nu} \\ &= \lim_{x\to a+} \frac{\varphi(x)}{\rho(x)} \left[\int_{(a,x]} \rho(t)\, d\nu + \rho(x) \int_{(x,b)} d\nu \right] \\ &\le \lim_{x\to a+} \left[\int_{(a,x]} \varphi(t)\, d\nu + \varphi(x) \int_{(x,b)} d\nu \right] \\ &\le \int_{(a,x_{K_-+1}]} \varphi(t)\, d\nu + \sup_{a<x\le x_{K_-+1}} \varphi(x) \int_{(x,b)} d\nu \\ &=: A_1 + A_2. \end{aligned} \tag{4.2.84}$$

On the other hand, since ρ is increasing on I,

$$\begin{aligned} A &\ge \limsup_{x\to a+} \varphi(x) \int_{(x,b)} d\nu \ge \varphi(x_{K_-}) \lim_{x\to a+} \int_{(x,x_{K_-+1}]} d\bar{\nu} \\ &= \varphi(x_{K_-}) \int_{(x_{K_-},x_{K_-+1}]} d\bar{\nu} =: A_0; \end{aligned} \tag{4.2.85}$$

note that the assumptions $K_- > -\infty$ and (4.2.56) imply that $\varphi(x_{K_-}) > 0$ - cf. Remarks 4.2.2 (i). By (4.2.83)–(4.2.85),

$$A_0 + N \le \text{RHS(4.2.59)} \le A_1 + A_2 + N. \tag{4.2.86}$$

Now, (4.2.82), (4.2.72),(4.2.71), (4.2.73) and (4.2.74) show that

$$N \lesssim \text{RHS}(4.2.79),$$

and, on using (4.2.75) and (4.2.76), we obtain that

$$N \lesssim \text{RHS}(4.2.80). \tag{4.2.87}$$

It is also clear that

$$A_1 = \int_{(x_{K_-}, x_{K_-+1}]} \varphi(t)\, d\nu \le \text{RHS}(4.2.80). \tag{4.2.88}$$

Moreover, taking into account (4.2.34), we observe that

$$\begin{aligned} A_2 &\le \sup_{a<x\le x_{K_-+1}} \varphi(x) \int_{(x, x_{K_-+1}]} d\nu + \varphi(x_{K_-+1}) \int_{(x_{K_-+1}, b)} d\nu \\ &=: A_{21} + A_{22}. \end{aligned} \tag{4.2.89}$$

By (4.2.62),

$$A_{22} \le N_2 \lesssim N. \tag{4.2.90}$$

Since φ is non-decreasing on I,

$$A_{21} \le \sup_{a<x\le x_{K_-+1}} \int_{(x, x_{K_-+1}]} \varphi(t)\, d\nu = \int_{(x_{K_-}, x_{K_-+1}]} \varphi(t)\, d\nu.$$

Together with (4.2.80), this shows that

$$A_{21} \lesssim N. \tag{4.2.91}$$

Combining estimates (4.2.86)–(4.2.91), we arrive at

$$\text{RHS}(4.2.59) \lesssim \text{RHS}(4.2.80). \tag{4.2.92}$$

On the other hand, by (4.2.71), (4.2.85), (4.2.81) and (4.2.82),

$$A_0 + N \approx \left\| \left(\frac{\varphi}{\rho}\right)(x_k) \int_{(x_{k-1}, x_k]} \rho(t)\, d\nu \right\|_{\ell^q(\mathcal{K}^+)} + \left\| \varphi(x_{k-1}) \int_{(x_{k-1}, x_k]} d\nu \right\|_{\ell^q(\mathcal{K}^+)},$$

so that

$$A_0 + N \approx \text{RHS}(4.2.78).$$

Consequently, applying (4.2.73), (4.2.74), (4.2.75) and (4.2.76), we arrive at

$$A_0 + N \approx \text{RHS}(4.2.80). \tag{4.2.93}$$

Combining estimates (4.2.86), (4.2.92) and (4.2.93), we obtain that RHS(4.2.59) $\approx$ RHS(4.2.80), and thus the second estimate in (4.2.57) follows from (4.2.59) and the definition of RHS(4.2.80).

(iii) Assume now that

$$K_- = -\infty \quad \text{and} \quad K_+ < +\infty. \tag{4.2.94}$$

(Note that $\mathcal{K} \subsetneqq \mathcal{K}^+$ now.) Then

$$\text{RHS}(4.2.59) = N + B. \tag{4.2.95}$$

Using (4.1.32), (4.2.58) and (4.2.33), we see that

$$\begin{aligned} B &= \lim_{x\to b-} \Big(\frac{\varphi}{\rho}\Big)(x)\Big[\int_{(a,x]} \rho(t)\,d\nu + \varphi(x)\int_{(x,b)} d\nu\Big] \\ &= \lim_{x\to b-} \Big[\Big(\frac{\varphi}{\rho}\Big)(x)\int_{(a,x_{K_+-1}]} \rho(t)\,d\nu + \Big(\frac{\varphi}{\rho}\Big)(x)\int_{(x_{K_+-1},x]} \rho(t)\,d\nu \\ &\quad + \varphi(x)\int_{(x,b)} d\nu\Big] \\ &\le \Big(\frac{\varphi}{\rho}\Big)(x_{K_+-1})\int_{(a,x_{K_+-1}]} \rho(t)\,d\nu \\ &\quad + \limsup_{x\to b-} \Big(\frac{\varphi}{\rho}\Big)(x)\int_{(x_{K_+-1},x]} \rho(t)\,d\nu + \limsup_{x\to b-} \varphi(x)\int_{(x,b)} d\nu \\ &=: B_1 + B_2 + B_3. \end{aligned} \tag{4.2.96}$$

By (4.2.63), (4.2.64), (4.2.9), (1.3.3) of Lemma 1.3.4 (with $L = K_- + 1$ and $M = K_+ - 1$) and (4.2.71),

$$\begin{aligned} B_1 &= \Big(\frac{\varphi}{\rho}\Big)(x_{K_+-1})\int_{(a,x_{K_+-1}]} \rho(t)\,d\nu = \tau_{K_+-1}\sum_{m=K_-+1}^{K_+-1} a_m \\ &\le \Big\|\tau_k \sum_{m=K_-+1}^{k} a_m\Big\|_{\ell^q(\mathcal{K})} \approx \|\tau_k a_k\|_{\ell^q(\mathcal{K})} \\ &= \Big\|\Big(\frac{\varphi}{\rho}\Big)(x_k)\int_{(x_{k-1},x_k]} \rho(t)\,d\nu\Big\|_{\ell^q(\mathcal{K})} \lesssim N. \end{aligned} \tag{4.2.97}$$

Moreover, by (4.2.33),

$$\begin{aligned} B_2 &\le \lim_{x\to b-} \int_{(x_{K_+-1},x]} \varphi(t)\,d\nu = \int_{(x_{K_+-1},b)} \varphi(t)\,d\nu \\ &= \int_{(x_{K_+-1},x_{K_+}]} \varphi(t)\,d\bar{\nu} =: B^*, \end{aligned} \tag{4.2.98}$$

and, by (4.2.34),

$$B_3 \le B^*. \tag{4.2.99}$$

Consequently, (cf. (4.2.96)–(4.2.99))

$$N + B \lesssim N + B^*. \tag{4.2.100}$$

Assume now that

$$K_+ \in \mathcal{Z}_1. \tag{4.2.101}$$

Then, by (4.1.7),

$$B^* \approx \varphi(x_{K_+-1}) \int_{(x_{K_+-1}, x_{K_+}]} d\bar{\nu}.$$

Together with (4.2.71) and (4.2.100), this shows that

$$N + B \lesssim N.$$

The last estimate and (4.2.95) imply that

$$\text{RHS}(4.2.59) \approx N. \tag{4.2.102}$$

Using (4.2.94), (4.2.101), (4.2.72) and (4.2.71), we obtain that

$$N \gtrsim \left\| \left(\frac{\varphi}{\rho}\right)(x_k) \int_{(x_{k-1}, x_k]} \rho(t)\, d\bar{\nu} \right\|_{\ell^q(\mathcal{Z}_2)} + \left\| \varphi(x_{k-1}) \int_{(x_{k-1}, x_k]} d\bar{\nu} \right\|_{\ell^q(\mathcal{Z}_1)}.$$

By (4.2.71), (4.2.73) and (4.2.74), the reverse estimate holds as well. Consequently, we have arrived at (4.2.79), which, together with (4.2.75), (4.2.76) and (4.2.72), implies (4.2.80). The second estimate in (4.2.57) follows from (4.2.80), (4.2.102) and (4.2.59).

Assume now that

$$K_+ \in \mathcal{Z}_2. \tag{4.2.103}$$

Then, by (4.1.32), (4.2.58), (4.1.8) and the definition of B^* in (4.2.98),

$$\begin{aligned} B &\geq \limsup_{x \to b-} \left(\frac{\varphi}{\rho}\right)(x) \int_{(x_{K_+-1}, x]} \rho(t)\, d\nu \\ &\approx \limsup_{x \to b-} \int_{(x_{K_+-1}, x]} \varphi(t)\, d\nu \\ &= B^*. \end{aligned} \tag{4.2.104}$$

Using (4.2.95), (4.2.100) and (4.2.104), we arrive at

$$\text{RHS}(4.2.59) \approx N + B^*, \tag{4.2.105}$$

which, together with (4.2.71), (4.2.103), (4.2.94), the definition of B^* and (4.1.8), shows that

$$\text{RHS}(4.2.59) \approx \left\| \left(\frac{\varphi}{\rho}\right)(x_k) \int_{(x_{k-1}, x_k]} \rho(t)\, d\bar{\nu} \right\|_{\ell^q(\mathcal{K}+)} + \left\| \varphi(x_{k-1}) \int_{(x_{k-1}, x_k]} d\bar{\nu} \right\|_{\ell^q(\mathcal{K}+)};$$

recall that $\{K_-+2,...,K_+\} = \mathcal{K}^+$ since $K_- = -\infty$. Hence, we obtain the second estimate in (4.2.57) just as in part (i).

(iv) The final case

$$K_- > -\infty \quad \text{and} \quad K_+ < +\infty$$

is treated by arguments similar to those used in the other cases above. □

Similar arguments to those used to prove Theorem 4.2.5 can be utilized to prove the following theorem, which generalizes the result mentioned in Remark 4.2.4 (i).

Theorem 4.2.7 *Let $I = (a,b) \subseteq \mathbb{R}$, $q \in (0,+\infty]$, $\mu \in \mathcal{B}^+(I)$ and let $w \in \mathcal{W}(I,\mu)$. Put*

$$\varphi(x) := \|\min\{\rho(\cdot),\rho(x)\}\|_{q,w,I,\mu}, \quad x \in I,$$

and assume that

$$0 < \varphi(\bar{x}) < +\infty \text{ for some } \bar{x} \in I.$$

Let $\{x_k\}_{k=K_-}^{K_+} \in CS(\varphi,\rho,I,\alpha)$ with $\alpha > 2^{1/q}$. Then, for all $\nu \in \mathcal{B}^+(I)$ and $g \in \mathcal{M}^+(I,\nu)$,

$$\begin{aligned} &\| \|\min\{\rho(\cdot),\rho(x)\}\|_{\infty,g,I,\nu}\|_{q,w,I,\mu} \\ &\approx \left\| \frac{\varphi(x_k)}{\rho(x_k)} \|\min\{\rho(\cdot),\rho(x_k)\}\|_{\infty,g,I,\nu} \right\|_{\ell^q(\mathcal{K}_-^+)} \\ &\approx \| \|\varphi\|_{\infty,g,(x_{k-1},x_k],\bar{\nu}}\|_{\ell^q(\mathcal{K}^+)}, \end{aligned} \tag{4.2.106}$$

where $\bar{\nu}$ is the extension of ν by zero in $\mathbb{R}\setminus I$ (cf. page 2).

Remark 4.2.8 Again, it follows from the proof of Theorem 4.2.7 that the second equivalence in (4.2.106) holds for any $\varphi \in Q_\rho(I)$, $0 \not\equiv \varphi \not\equiv +\infty$, such that $\{x_k\}_{k=K_-}^{K_+} \in CS(\varphi,\rho,I,\alpha)$ with $\alpha > 1$.

In Chapter 5 we shall need the following result.

Lemma 4.2.9 *Let $I = (a,b) \subseteq \mathbb{R}$, $p \in (0,+\infty]$, $\rho \in Ads(I)$ and $f_1, f_2 \in Q_\rho(I)$. Assume that $\{x_k^i\}_{k=K_-^i}^{K_+^i} \in CS(f_i,\rho,I,\alpha_i)$, $\alpha_i > 1$, and $\mathcal{K}_-^{i+} := \{k \in \mathbb{Z};\ K_-^i \le k \le K_+^i\}$, $i = 1,2$. Then*

$$\left\| \left(\frac{f_1}{\rho}\right)(x_k^1)\, f_2(x_k^1) \right\|_{\ell^p(\mathcal{K}_-^{1+})} \approx \left\| \left(\frac{f_1}{\rho}\right)(x_k^2)\, f_2(x_k^2) \right\|_{\ell^p(\mathcal{K}_-^{2+})}. \tag{4.2.107}$$

Proof. We may clearly assume that $0 \not\equiv f_i \not\equiv +\infty$, $i = 1,2$. We shall only prove the inequality

$$\text{LHS(4.2.107)} \lesssim \text{RHS(4.2.107)} \tag{4.2.108}$$

since the reverse estimate follows by interchanging the roles of f_1 and f_2.

Define

$$\mathcal{K}^{1+}_{j-} := \{k \in \mathcal{K}^{1+}_{-};\ x^2_{j-1} \le x^1_k \le x^2_j\}, \quad j \in \mathcal{K}^{2+}, \tag{4.2.109}$$

and set

$$\mathcal{K}^1_0 := \bigcup_{j \in \mathcal{K}^{2+}} \mathcal{K}^{1+}_{j-} \tag{4.2.110}$$

(note that at most two sets appearing on the right-hand side of (4.2.110) may have a nonempty intersection).

We shall distinguish two cases:

$$\mathcal{K}^{1+}_{-} = \mathcal{K}^1_0 \tag{4.2.111}$$

and

$$\mathcal{K}^{1+}_{-} \neq \mathcal{K}^1_0. \tag{4.2.112}$$

Note that case (4.2.112) appears if and only if

$$K^1_+ \notin \mathcal{K}^1_0 \quad \text{or} \quad K^1_- \notin \mathcal{K}^1_0,$$

which, in turn, correspond to the conditions

$$J^1_+ < +\infty = J^2_+ \quad \text{or} \quad J^1_- > -\infty = J^2_-, \tag{4.2.113}$$

respectively.

By Lemma 3.3.1, there is a decomposition of the set $\mathcal{K}^{2+}$,

$$\mathcal{K}^{2+} = \mathcal{Z}^2_1 \cup \mathcal{Z}^2_2, \qquad \mathcal{Z}^2_1 \cap \mathcal{Z}^2_2 = \emptyset, \tag{4.2.114}$$

such that

$$f_2(x) \approx f_2(x^2_j) \ \text{ for all } x \in [x^2_{j-1}, x^2_j] \ \text{ and every } j \in \mathcal{Z}^2_1, \tag{4.2.115}$$

and

$$\Big(\frac{\rho}{f_2}\Big)(x) \approx \Big(\frac{\rho}{f_2}\Big)(x^2_j) \ \text{ for all } x \in [x^2_{j-1}, x^2_j] \ \text{ and every } j \in \mathcal{Z}^2_2. \tag{4.2.116}$$

(i) First consider the case (4.2.111). Then, on using also (4.2.110) and (4.2.114), we arrive at

$$\begin{aligned}
\text{LHS}(4.2.107) &= \Big\| \Big(\frac{f_1}{\rho}\Big)(x^1_k)\, f_2(x^1_k) \Big\|_{\ell^p(\mathcal{K}^1_0)} \\
&\lesssim \Big\| \Big\| \Big(\frac{f_1}{\rho}\Big)(x^1_k)\, f_2(x^1_k) \Big\|_{\ell^p(\mathcal{K}^{1+}_{j-})} \Big\|_{\ell^p(\mathcal{K}^{2+})} \\
&\approx \Big\| \Big\| \Big(\frac{f_1}{\rho}\Big)(x^1_k)\, f_2(x^1_k) \Big\|_{\ell^p(\mathcal{K}^{1+}_{j-})} \Big\|_{\ell^p(\mathcal{Z}^2_1)} \\
&\quad + \Big\| \Big\| \Big(\frac{f_1}{\rho}\Big)(x^1_k)\, f_2(x^1_k) \Big\|_{\ell^p(\mathcal{K}^{1+}_{j-})} \Big\|_{\ell^p(\mathcal{Z}^2_2)}.
\end{aligned} \tag{4.2.117}$$

If $j \in \mathcal{Z}_1^2$, then, applying (4.2.115), Lemma 1.3.3 when $0 < p < +\infty$ (the case $p = +\infty$ is trivial), and the fact that $f_1/\rho \in \mathcal{M}^+(I;\downarrow)$, we obtain

$$\left\| \left(\frac{f_1}{\rho}\right)(x_k^1)\, f_2(x_k^1) \right\|_{\ell^p(\mathcal{K}_{j-}^{1+})} \approx f_2(x_{j-1}^2) \left\| \left(\frac{f_1}{\rho}\right)(x_k^1) \right\|_{\ell^p(\mathcal{K}_{j-}^{1+})} \lesssim \left(\frac{f_1}{\rho}\right)(x_{j-1}^2)\, f_2(x_{j-1}^2).$$

Similarly, if $j \in \mathcal{Z}_2^2$, then (4.2.116), Lemma 1.3.3 when $0 < p < +\infty$ (the case $p = +\infty$ is again trivial), and the fact that $f_1 \in \mathcal{M}^+(I;\uparrow)$ imply that

$$\left\| \left(\frac{f_1}{\rho}\right)(x_k^1)\, f_2(x_k^1) \right\|_{\ell^p(\mathcal{K}_{j-}^{1+})} \approx \left(\frac{f_2}{\rho}\right)(x_j^2) \left\| f_1(x_k^1) \right\|_{\ell^p(\mathcal{K}_{j-}^{1+})} \lesssim \left(\frac{f_2}{\rho}\right)(x_j^2) f_1(x_j^2).$$

Combining the previous two estimates with (4.2.117), we arrive at

$$\mathrm{LHS}(4.2.107) \lesssim \left\| \left(\frac{f_1}{\rho}\right)(x_{j-1}^2)\, f_2(x_{j-1}^2) \right\|_{\ell^p(\mathcal{Z}_1^2)} + \left\| \left(\frac{f_1}{\rho}\right)(x_j^2)\, f_2(x_j^2) \right\|_{\ell^p(\mathcal{Z}_2^2)}$$

$$\lesssim \mathrm{RHS}(4.2.107),$$

which shows that (4.2.108) holds when (4.2.111) is satisfied.

(ii) Consider now the case (4.2.112). We shall look at the most involved situation, when

$$J_+^1 < +\infty = J_+^2 \quad \text{and} \quad J_-^1 > -\infty = J_-^2;$$

proofs for the other possibilities in (4.2.113) are similar, but simpler, and are left to the reader. Then

$$\mathcal{K}_-^{1+} = \mathcal{K}_0^1 \cup \{K_+^1\} \cup \{K_-^1\}. \tag{4.2.118}$$

Hence,

$$\mathrm{LHS}(4.2.107) \lesssim \left\| \left(\frac{f_1}{\rho}\right)(x_k^1)\, f_2(x_k^1) \right\|_{\ell^p(\mathcal{K}_0^1)} + \left(\frac{f_1}{\rho}\right)(x_{K_+^1}^1)\, f_2(x_{K_+^1}^1) + \left(\frac{f_1}{\rho}\right)(x_{K_-^1}^1)\, f_2(x_{K_-^1}^1). \tag{4.2.119}$$

(ii-1) The inequality $J_+^1 < +\infty$ implies that

$$K_+^1 = J_+^1 + 1, \quad x_{K_+^1}^1 = b,$$

and, by Lemma 3.2.4,

(4.2.120) $$\Big(\frac{f_1}{\rho}\Big)(b) > 0 \qquad \text{or} \qquad f_1(b) < +\infty.$$

Moreover, the equality $J^2_+ = +\infty$ and Lemma 3.2.4 show that

$$\lim_{k\to+\infty} x^2_k = b,$$

(4.2.121) $$\lim_{k\to+\infty} f_2(x^2_k) = +\infty \quad \text{and} \quad \lim_{k\to+\infty} \frac{\rho(x^2_k)}{f_2(x^2_k)} = +\infty.$$

The first options in (4.2.120) and (4.2.121) give

$$\lim_{k\to+\infty} \frac{f_1(x^2_k)f_2(x^2_k)}{\rho(x^2_k)} = +\infty.$$

Thus, RHS (4.2.107) $= +\infty$, *which means that estimate* (4.2.108) *is satisfied.*
The other option in (4.2.120), namely $f_1(b) < +\infty$, and (4.2.121) give

$$\frac{f_1(b)f_2(b)}{\rho(b)} = 0$$

and hence, *the second term on the right-hand side of* (4.2.119) *is absent.*

(ii-2) The inequality $J^1_- > -\infty$ implies that

$$K^1_- = J^1_- - 1, \quad x^1_{K^1_-} = a,$$

and, by Lemma 3.2.4,

(4.2.122) $$\Big(\frac{f_1}{\rho}\Big)(a) < +\infty \qquad \text{or} \qquad f_1(a) > 0.$$

Moreover, the equality $J^2_- = -\infty$ and Lemma 3.2.4 show that

$$\lim_{k\to-\infty} x^2_k = a,$$

(4.2.123) $$\lim_{k\to-\infty} f_2(x^2_k) = 0 \quad \text{and} \quad \lim_{k\to-\infty} \frac{f_2(x^2_k)}{\rho(x^2_k)} = +\infty.$$

The first options in (4.2.122) and (4.2.123) imply that

$$\frac{f_1(a)f_2(a)}{\rho(a)} = 0$$

and hence, *the third term on the right-hand side of* (4.2.119) *is absent.*
The other option in (4.2.122), namely, $f_1(a) > 0$, and (4.2.123) give

$$\lim_{k\to-\infty} \frac{f_1(x^2_k)f_2(x^2_k)}{\rho(x^2_k)} = +\infty.$$

Thus, RHS(4.2.107)$= +\infty$, *which means that estimate* (4.2.108) *is satisfied.*

To finish the proof, it remains to show that the first term on the right-hand side of (4.2.119) is dominated by a constant multiple of RHS(4.2.107). However, this can be done in the same way as in part (i). $\square$

In the next chapter we shall also need the following corollary involving three functions $f_1, f_2, F \in Q_\rho(I)$.

Corollary 4.2.10 *Let* $I = (a,b) \subseteq \mathbb{R}$, $p \in (0,+\infty]$, $\rho \in Ads(I)$, $f_1, f_2, F \in Q_\rho(I)$ *and* $f_1 \approx f_2$ *on* I. *Assume that* $\{x_k^i\}_{k=K_-^i}^{K_+^i} \in CS(f_i, \rho, I, \alpha_i)$, $\alpha_i > 1$, *and* $\mathcal{K}_-^{i+} := \{k \in \mathbb{Z};\ K_-^i \le k \le K_+^i\}$, $i = 1,2$. *Then*

$$\left\| \left(\frac{f_1}{\rho}\right)(x_k^1)\, F(x_k^1) \right\|_{\ell^p(\mathcal{K}_-^{1+})} \approx \left\| \left(\frac{f_2}{\rho}\right)(x_k^2)\, F(x_k^2) \right\|_{\ell^p(\mathcal{K}_-^{2+})}. \tag{4.2.124}$$

Proof. Let

$$\{y_k\}_{k=K_-}^{K_+} \in CS(F, \rho, I, \alpha), \quad \alpha > 1,$$

and

$$\mathcal{K}_-^{+} := \{k \in \mathbb{Z};\ K_- \le k \le K_+\}.$$

By Lemma 4.2.9, and since $f_1 \approx f_2$, we have that

$$\begin{aligned}
\left\| \left(\frac{f_1}{\rho}\right)(x_k^1)\, F(x_k^1) \right\|_{\ell^p(\mathcal{K}_-^{1+})} &\approx \left\| \left(\frac{f_1}{\rho}\right)(y_k)\, F(y_k) \right\|_{\ell^p(\mathcal{K}_-^{+})} \\
&\approx \left\| \left(\frac{f_2}{\rho}\right)(y_k)\, F(y_k) \right\|_{\ell^p(\mathcal{K}_-^{+})} \\
&\approx \left\| \left(\frac{f_2}{\rho}\right)(x_k^2)\, F(x_k^2) \right\|_{\ell^p(\mathcal{K}_-^{2+})}.
\end{aligned}$$

$\square$

Chapter 5

Weighted inequalities for ρ-quasiconcave operators

5.1 The inequality $\left\| \sup_{a<t\le x} \rho(t) \int_t^b g(s)\, ds \right\|_{q,w,I} \lesssim \|g\|_{p,v,I}$

In this section we apply our previous results to characterize the validity of the exhibited weighted inequality involving ρ-quasiconcave operators. The reverse inequality will be treated in the next section, and in the remaining two sections of this chapter, another weighted inequality and the associated reverse inequality will be investigated. The method we employ consists of several steps. Firstly, we discretize both sides of the inequality in question. Secondly, we solve our problem locally (which represents an easier task) to get a discrete characterization of the original inequality. Finally, we apply an antidiscretization to convert the discrete characterization to a continuous one.

Theorem 5.1.1 *Let $I = (a,b) \subseteq \mathbb{R}$, $\rho \in Ads(I)$ and $w, v \in \mathcal{W}(I)$. Assume that $1 \le p \le +\infty$ and $0 < q \le +\infty$. Then*

$$\left\| \sup_{a<t\le x} \rho(t) \int_t^b g(s)\, ds \right\|_{q,w,I} \lesssim \|g\|_{p,v,I} \quad \text{for all } g \in \mathcal{M}^+(I), \tag{5.1.1}$$

if and only if one of the following is satisfied:
Case A: *$p \le q$ and*

$$\left\| \left\| \min\left\{\frac{\rho(\cdot)}{\rho(x)}, 1\right\} \right\|_{q,w,I} \sup_{a<t\le x} \rho(t) \|v^{-1}\|_{p',(t,b)} \right\|_{\infty,I} < +\infty; \tag{5.1.2}$$

Case B: *$q < p$, and*

$$\left\| \left\| \min\left\{\frac{\rho(\cdot)}{\rho(x)}, 1\right\} \right\|_{q,w,I}^{1-q/r} w^{q/r}(x) \sup_{a<t\le x} \rho(t) \|v^{-1}\|_{p',(t,b)} \right\|_{r,I} < +\infty, \tag{5.1.3}$$

where $1/r := 1/q - 1/p$.

Remark 5.1.2 Let all the assumptions of Theorem 5.1.1 be satisfied. Assume that $q = +\infty$ and $\varphi \not\equiv +\infty$, where φ is given by

$$\varphi(x) := \| \min\{\rho(\cdot), \rho(x)\}\|_{q,w,I}, \; x \in I. \tag{5.1.4}$$

Then condition (5.1.2) can be rewritten as

$$\| \sup_{a<t\le x} \rho(t) \|v^{-1}\|_{p',(t,b)} \|_{\infty,w,I} < +\infty. \tag{5.1.5}$$

Indeed, since the function

$$h(x) := \sup_{a<t\le x} \rho(t) \|v^{-1}\|_{p',(t,b)}, \quad x \in I,$$

belongs to $Q_\rho(I)$ (cf. Lemma 2.2.9), we have from (5.1.2) and Lemma 2.2.14 that

$$\mathrm{LHS}(5.1.5) = \|h\|_{\infty,w,I} = \|h\|_{\infty,\varphi/\rho,I} = \mathrm{LHS}(5.1.2).$$

The proof of Theorem 5.1.1 is dependent on two lemmas.

Lemma 5.1.3 *Let* $I = (a,b) \subseteq \mathbb{R}$, $\rho \in Ads(I)$ *and* $w, v \in \mathcal{W}(I)$. *Assume that* $1 \le p \le +\infty$ *and* $0 < q \le +\infty$. *Let*

$$\varphi(x) := \| \min\{\rho(\cdot), \rho(x)\}\|_{q,w,I} < +\infty \quad \text{for all} \quad x \in I, \tag{5.1.6}$$

and

$$\{x_k\}_{k=K_-}^{K_+} \in CS(\varphi, \rho, I, \alpha) \text{ with } \alpha > 2^{1/q}. \tag{5.1.7}$$

Define [1]

$$A_k := \sup_{h \in \mathcal{M}^+(I_k)} \Big(\sup_{t \in I_k} \varphi(t) \int_t^{x_k} h(y)\, dy \Big) \Big/ \|hv\|_{p,I_k}, \quad k \in \mathcal{K}^+, \tag{5.1.8}$$

where $I_k := (x_{k-1}, x_k]$. *Then inequality* (5.1.1) *holds if and only if*

$$\|\{A_k\}\|_{\ell^r(\mathcal{K}^+)} < +\infty, \tag{5.1.9}$$

where $\frac{1}{r} := \left(\frac{1}{q} - \frac{1}{p}\right)_+$.

Proof. By Theorem 4.2.1,

$$\mathrm{LHS}(5.1.1) \approx \Big\| \sup_{t \in I_k} \varphi(t) \int_t^{x_k} g(s)\, ds \Big\|_{\ell^q(\mathcal{K}^+)}. \tag{5.1.10}$$

[1] Note that in (5.1.8) we use the convention that $0/0 = 0$ and $(+\infty)/(+\infty) = 0$ - cf. Convention 1.1.1 (i).

Since

$$\text{RHS}(5.1.1) = \| \, \|gv\|_{p,I_k} \|_{\ell^p(\mathcal{K}^+)}, \tag{5.1.11}$$

inequality (5.1.1) can be rewritten as

$$\left\| \sup_{t\in I_k} \varphi(t) \int_t^{x_k} g(s)\,ds \right\|_{\ell^q(\mathcal{K}^+)} \lesssim \| \, \|gv\|_{p,I_k} \|_{\ell^p(\mathcal{K}^+)} \quad \text{for all } g \in \mathcal{M}^+(I). \tag{5.1.12}$$

To find a sufficient condition for the validity of inequality (5.1.12), we apply locally (that is, for any $k \in \mathcal{K}^+$) the Hardy-type inequality

$$\sup_{t\in I_k} \varphi(t) \int_t^{x_k} h(s)\,ds \le A_k \, \|hv\|_{p,I_k} \quad \text{for all } h \in \mathcal{M}^+(I_k), \tag{5.1.13}$$

where the best possible constant A_k is that defined in (5.1.8).

If $p \le q < +\infty$, then, by (5.1.13) and since $q/p \ge 1$, we obtain

$$\begin{aligned}
[\text{LHS}(5.1.12)]^q &= \sum_{k\in\mathcal{K}^+} \left[\sup_{t\in I_k} \varphi(t) \int_t^{x_k} g(s)\,ds \right]^q \\
&\le \sum_{k\in\mathcal{K}^+} [A_k \, \|gv\|_{p,I_k}]^q \\
&\le \Big(\sup_{k\in\mathcal{K}^+} A_k \Big)^q \Big(\sum_{k\in\mathcal{K}^+} \int_{I_k} g^p(s)\, v^p(s)\,ds \Big)^{q/p} \\
&= \|\{A_k\}\|^q_{\ell^\infty(\mathcal{K}^+)} \, \|gv\|^q_{p,I}.
\end{aligned}$$

If $q = +\infty$, then, by (5.1.13),

$$\begin{aligned}
\text{LHS}(5.1.12) &= \sup_{k\in\mathcal{K}^+} \left[\sup_{t\in I_k} \varphi(t) \int_t^{x_k} g(s)\,ds \right] \\
&\le \sup_{k\in\mathcal{K}^+} A_k \, \|gv\|_{p,I_k} \\
&\le \|\{A_k\}\|_{\ell^\infty(\mathcal{K}^+)} \, \|gv\|_{p,I}.
\end{aligned}$$

In the case that $q < p$, Hölder's inequality for sums (with the exponents r/q and p/q) applied to (5.1.13) gives

$$\begin{aligned}
[\text{LHS}(5.1.12)]^q &= \sum_{k\in\mathcal{K}^+} \left[\sup_{t\in I_k} \varphi(t) \int_t^{x_k} g(s)\,ds \right]^q \\
&\le \sum_{k\in\mathcal{K}^+} [A_k \, \|gv\|_{p,I_k}]^q \\
&\le \Big(\sum_{k\in\mathcal{K}^+} A_k^r \Big)^{q/r} \Big(\sum_{k\in\mathcal{K}^+} \|gv\|^p_{p,I_k} \Big)^{q/p} \\
&= \|\{A_k\}\|^q_{\ell^r(\mathcal{K}^+)} \|gv\|^q_{p,I}.
\end{aligned}$$

Therefore in all cases, (5.1.12), and hence (5.1.1), holds if condition (5.1.9) is satisfied. (Note that inequality (5.1.13) holds with a finite constant $A_k, k \in \mathcal{K}^+$, if (5.1.9) is fulfilled.)

Now, we prove that condition (5.1.9) is also necessary for the validity of inequality (5.1.12) (and so for (5.1.1), too). By (5.1.8), there are $h_k \in \mathcal{M}^+(I_k)$, $k \in \mathcal{K}^+$, such that

$$\|h_k\|_{p,v,I_k} = 1 \tag{5.1.14}$$

and

$$\sup_{t\in I_k} \varphi(t) \int_t^{x_k} h_k(s)\,ds \geq \frac{1}{2} A_k. \tag{5.1.15}$$

Define g_k, $k \in \mathcal{K}^+$, as the extension of h_k by 0 to the whole interval I and put

$$g = \sum_{k\in\mathcal{K}^+} a_k g_k, \tag{5.1.16}$$

where $\{a_k\}_{k\in\mathcal{K}^+}$ is any sequence of positive numbers. Using the fact that

$$\operatorname{supp} g_k \subset [x_{k-1}, x_k] \tag{5.1.17}$$

and (5.1.14), we obtain

$$\begin{aligned} \text{RHS(5.1.12)} &= \Big\| \Big\| \Big(\sum_{m\in\mathcal{K}^+} a_m g_m \Big) v \Big\|_{p,I_k} \Big\|_{\ell^p(\mathcal{K}^+)} \\ &= \| a_k \| g_k v \|_{p,I_k} \|_{\ell^p(\mathcal{K}^+)} \\ &= \| \{a_k\} \|_{\ell^p(\mathcal{K}^+)}. \end{aligned} \tag{5.1.18}$$

Moreover, making use of (5.1.17) and (5.1.15), we arrive at

$$\begin{aligned} \text{LHS(5.1.12)} &= \Big\| \sup_{t\in I_k} \varphi(t) \int_t^{x_k} \Big(\sum_{m\in\mathcal{K}^+} a_m g_m(s) \Big)\,ds \Big\|_{\ell^q(\mathcal{K}^+)} \\ &= \Big\| a_k \sup_{t\in I_k} \varphi(t) \int_t^{x_k} g_k(s)\,ds \Big\|_{\ell^q(\mathcal{K}^+)} \\ &\geq \frac{1}{2} \| \{a_k A_k\} \|_{\ell^q(\mathcal{K}^+)}. \end{aligned} \tag{5.1.19}$$

Therefore, by (5.1.12), (5.1.19) and (5.1.18),

$$\| \{a_k A_k\} \|_{\ell^q(\mathcal{K}^+)} \lesssim \| \{a_k\} \|_{\ell^p(\mathcal{K}^+)} \tag{5.1.20}$$

for all sequences $\{a_k\}_{k\in\mathcal{K}^+}$ of positive numbers. Consequently, by Lemma 1.4.1 when $p \leq q$ and Lemma 1.4.2 when $q < p$, (5.1.9) is satisfied and the lemma is proved. □

Lemma 5.1.4 *Suppose that the conditions of* Lemma 5.1.3 *are satisfied. Then*

$$A_k \approx \tilde{A}_k := \sup_{t\in I_k} \varphi(t)\|v^{-1}\|_{p',(t,x_k)} \quad \textit{for all } k \in \mathcal{K}^+, \tag{5.1.21}$$

and so, (5.1.1) *is satisfied if and only if*

$$\|\{\tilde{A}_k\}\|_{\ell^r(\mathcal{K}^+)} < +\infty, \tag{5.1.22}$$

where $\frac{1}{r} := \left(\frac{1}{q} - \frac{1}{p}\right)_+$.

Proof. The lemma is an immediate consequence of the well-known fact (see [OK]) that for $1 \le p \le +\infty$,

$$A_k \approx \sup_{t\in I_k} \|\varphi\|_{\infty,(x_{k-1},t)} \|v^{-1}\|_{p',(t,x_k)} = \tilde{A}_k,$$

the equality being a consequence of $\varphi \in \mathcal{M}^+(I,\uparrow) \cap C(I)$. The rest follows from Lemma 5.1.3. □

Proof of Theorem 5.1.1. The function φ given by (5.1.4) belongs to $Q_\rho(I)$. Suppose first that it is such that

$$\varphi(x) = +\infty \quad \text{for some} \quad x \in I. \tag{5.1.23}$$

Then, by Remark 4.1.2, $\|f\|_{q,w,I} = +\infty$ for any $f \in Q_\rho(I)$, $f \not\equiv 0$. Since the function

$$f(x) := \sup_{a<t\le x} \rho(t) \int_t^b g(s)\, ds, \quad \text{where } g \in \mathcal{M}^+(I), \tag{5.1.24}$$

belongs to $Q_\rho(I)$ (cf. Corollary 2.2.10), and $f \not\equiv 0$ if $g \neq 0$ a.e. in I, it follows that LHS(5.1.1) $= +\infty$ for any $g \in \mathcal{M}^+(I)$ such that $g \neq 0$ a.e. in I. Consequently, (5.1.1) holds if and only if

$$\|g\|_{p,v,I} = +\infty \quad \text{for any } g \in \mathcal{M}^+(I) \text{ satisfying } g \neq 0 \text{ a.e. in } I. \tag{5.1.25}$$

However, (5.1.25) is equivalent to the assertion that

$$v = +\infty \text{ a.e. in } I.$$

Since this holds if and only if

$$\|v^{-1}\|_{p',(t,b)} = 0 \quad \text{for all } t \in I,$$

on using Convention 1.1.1 (i), we have that (5.1.1) holds if and only if LHS(5.1.2) $= 0$. But in our situation, this last condition is equivalent to the statement that LHS(5.1.2) $< +\infty$, which proves Theorem 5.1.1 in Case A when (5.1.23) is satisfied. The same argument applies to (5.1.3).

We may therefore assume (5.1.6), namely,

$$\varphi(x) < +\infty \quad \text{for all } x \in I.$$

Under this assumption, we proved in Lemma 5.1.4 that the inequality (5.1.1) is equivalent to (5.1.22) being satisfied. Therefore, it remains to verify that the "discrete condition" (5.1.22) is equivalent to the "continuous condition" (5.1.2) in Case A and to (5.1.3) in Case B of the theorem.

Case A: In this case, the discrete condition (5.1.22) becomes

$$\left\| \sup_{t \in I_k} \varphi(t) \, \|v^{-1}\|_{p',(t,x_k)} \right\|_{\ell^\infty(\mathcal{K}+)} < +\infty, \tag{5.1.26}$$

where

$$\{x_k\}_{k=K_-}^{K_+} \in CS(\varphi, \rho, I, \alpha) \quad \text{with } \alpha > 2^{1/q}.$$

It is necessary to consider separate cases within Case A.

Case A-1: $1 < p \le q \le +\infty$. Here we have $p' < +\infty$ and so, (5.1.26) is equivalent to

$$\left\| \sup_{t \in I_k} \varphi^{p'}(t) \int_t^{x_k} v^{-p'}(s)\, ds \right\|_{\ell^\infty(\mathcal{K}+)} < +\infty. \tag{5.1.27}$$

Even in this case it is convenient to distinguish two cases.

Case A-1-1: $1 < p \le q = +\infty$. In this case (5.1.4) implies that

$$\varphi^{p'}(x) = \| \min\{\rho^{p'}(\cdot), \rho^{p'}(x)\} w^{p'}(\cdot)\|_{\infty,I}, \quad x \in I,$$

i.e.,

$$\Phi(x) := \varphi^{p'}(x) = \| \min\{R(\cdot), R(x)\}\|_{\infty,W,I}, \quad x \in I,$$

where

$$R = \rho^{p'} \quad \text{and} \quad W = w^{p'}. \tag{5.1.28}$$

Moreover, by (5.1.6),

$$\Phi(x) < +\infty \qquad \text{for all } x \in I.$$

As also (cf. Remark 3.2.3 (ii))

$$\{x_k\}_{k=K_-}^{K_+} \in CS(\Phi, R, I, \alpha^{p'}), \quad \alpha > 2^{1/q}, \tag{5.1.29}$$

Theorem 4.2.1 (with $q = +\infty$) gives that (5.1.27) is equivalent to

$$\left\| \sup_{a < t \le x} R(t) \int_t^b v^{-p'}(s)\, ds \right\|_{\infty,W,I} < +\infty. \tag{5.1.30}$$

Since (5.1.30) can be rewritten as

$$\| \sup_{a<t\le x} R^{1/p'}(t)\, \|v^{-1}\|_{p',(t,b)} W^{1/p'}(x)\|_{\infty,I} < +\infty, \tag{5.1.31}$$

we arrive at

$$\| \sup_{a<t\le x} \rho(t)\, \|v^{-1}\|_{p',(t,b)} w(x)\|_{\infty,I} < +\infty,$$

which coincides with (5.1.5) and the desired results follows from Remark 5.1.2.

Case A-1-2: $1 < p \le q < +\infty$. Then, by (5.1.4),

$$\varphi^q(x) = \int_I \min\{\rho^q(t), \rho^q(x)\} w^q(t)\,dt, \quad x \in I, \tag{5.1.32}$$

and, on applying Theorem 2.4.3,

$$\begin{aligned} \varphi^{p'}(x) &= (\varphi^q(x))^{p'/q} \\ &\approx \int_I \min\{(\rho^q(t))^{p'/q} (\rho^q(x))^{p'/q}\} \Big(\frac{\varphi^q}{\rho^q}\Big)^{p'/q-1}(t)\, w^q(t)\,dt \\ &= \| \min\{\rho^{p'}(\cdot), \rho^{p'}(x)\}\|_{1,\omega,I} \qquad \text{for all } x \in I, \end{aligned} \tag{5.1.33}$$

where

$$\omega = \Big(\frac{\varphi^q}{\rho^q}\Big)^{p'/q-1} w^q. \tag{5.1.34}$$

Consequently, by Lemma 2.4.5,

$$\varphi^{p'}(x) \approx \| \min\{\rho^{p'}(\cdot), \rho^{p'}(x)\}\|_{\infty,W_\infty,I} \quad \text{for all } x \in I,$$

where

$$W_\infty = \frac{\varphi^{p'}}{\rho^{p'}}. \tag{5.1.35}$$

Hence,

$$\Phi(x) := \varphi^{p'}(x) \approx \| \min\{R(\cdot), R(x)\}\|_{\infty,W,I} =: \widetilde{\Phi}(x) \quad \text{for all } x \in I, \tag{5.1.36}$$

with

$$R = \rho^{p'} \quad \text{and} \quad W = W_\infty. \tag{5.1.37}$$

Moreover, by (5.1.6) and (5.1.36),

$$\Phi(x) < +\infty \quad \text{and} \quad \widetilde{\Phi}(x) < +\infty \qquad \text{for all } x \in I. \tag{5.1.38}$$

However, we cannot apply Theorem 4.2.1 directly to rewrite condition (5.1.27) in a continuous form, for while $\{x_k\}_{k=K_-}^{K_+}$ is a covering sequence of the function

φ (and so Φ as well), we do not know if it is a covering sequence of $\widetilde{\Phi}$. We shall prove that condition (5.1.27) is equivalent to

$$\left\| \sup_{t\in\widetilde{I}_k} \widetilde{\Phi}(t) \int_t^{\widetilde{x}_k} v^{-p'}(s)\,ds \right\|_{\ell^\infty(\widetilde{\mathcal{K}}^+)} < +\infty, \tag{5.1.39}$$

where

$$\{\widetilde{x}_k\}_{k=\widetilde{K}_-}^{\widetilde{K}_+} \in CS(\widetilde{\Phi}, R, I, \beta) \quad \text{with } \beta > 1, \tag{5.1.40}$$

$$\widetilde{\mathcal{K}}^+ := \{k \in \mathbb{Z} : \widetilde{K}_- < k \le \widetilde{K}_+\}, \tag{5.1.41}$$

and

$$\widetilde{I}_k := (\widetilde{x}_{k-1}, \widetilde{x}_k], \qquad k \in \widetilde{\mathcal{K}}^+. \tag{5.1.42}$$

An application of Theorem 4.2.1 (with $q = +\infty$) will then give that (5.1.39) is equivalent to (5.1.30). Since condition (5.1.30) can be rewritten as (5.1.31), (5.1.2) follows on using (5.1.37), (5.1.35) and (5.1.4).

Thus it remains to show that (5.1.27) and (5.1.39) are equivalent. Firstly, by Remark 4.2.2 (ii), condition (5.1.27) is equivalent to

$$\left\| \frac{\Phi(x_k)}{R(x_k)} \sup_{a<t\le x_k} R(t) \int_t^b v^{-p'}(s)\,ds \right\|_{\ell^\infty(\mathcal{K}_-^+)} < +\infty, \tag{5.1.43}$$

since $\Phi = \varphi^{p'}$, $R = \rho^{p'}$ and (5.1.29) holds. Secondly, let $\{\widetilde{x}_k\}_{k=\widetilde{K}_-}^{\widetilde{K}^+}$ be any covering sequence satisfying (5.1.40) and let $\widetilde{\mathcal{K}}^+$ and $\widetilde{I}_k$, $k \in \widetilde{\mathcal{K}}^+$, be given by (5.1.41) and (5.1.42). Then Corollary 4.2.10 (with functions $f_1 := \Phi, f_2 := \widetilde{\Phi}$, $\rho := R$ and $F(x) := \sup_{a<t\le x} R(t) \int_t^b v^{-p'}(s)\,ds,\ x \in I$) implies that condition (5.1.43) is equivalent to

$$\left\| \frac{\widetilde{\Phi}(\widetilde{x}_k)}{R(\widetilde{x}_k)} \sup_{a<t\le \widetilde{x}_k} R(t) \int_t^b v^{-p'}(s)\,ds \right\|_{\ell^\infty(\widetilde{\mathcal{K}}_-^+)} < +\infty. \tag{5.1.44}$$

Finally, again by Remark 4.2.2 (ii), the last condition is equivalent to (5.1.39), which completes the proof that conditions (5.1.27) and (5.1.39) are equivalent.[2]

Case A-2: $1 = p \le q \le +\infty$. Here we have $p' = +\infty$ and so (5.1.26) becomes

$$\| \sup_{t\in I_k} \varphi(t)\, \|v^{-1}\|_{\infty,(t,x_k)} \|_{\ell^\infty(\mathcal{K}^+)} < +\infty. \tag{5.1.45}$$

[2]Note that it would be possible to use the fact that (5.1.27) and (5.1.44) are equivalent and then to apply Theorem 4.2.1 to show that (5.1.44) is equivalent to the "continuous condition" (5.1.2). In our proof we have preferred to use the equivalence of conditions (5.1.27) and (5.1.39) because these two conditions have similar forms. However, some proofs of theorems in this chapter will correspond to the equivalence of conditions (5.1.27) and (5.1.44) since then proofs are slightly shorter.

Also in this case it is convenient to distinguish two cases.

Case A-2-1: $1 = p,\ q = +\infty$. Then

$$\varphi(x) = \|\min\{\rho(\cdot), \rho(x)\}\|_{\infty,w,I}, \quad x \in I,$$

and Theorem 4.2.3 (with $q = +\infty$) shows that (5.1.45) is equivalent to (5.1.5). Therefore, Remark 5.1.2 yields the desired result.

Case A-2-2: $1 = p \le q < +\infty$. Then, by (5.1.4), we have (5.1.32), and, by Theorem 2.4.3, this implies that

$$\begin{aligned} \varphi(x) = (\varphi^q(x))^{1/q} &\approx \int_I \min\{\rho(t), \rho(x)\} \left(\frac{\varphi^q}{\rho^q}\right)^{1/q-1}(t)\, w^q(t)\,dt \\ &= \|\min\{\rho(\cdot), \rho(x)\}\|_{1,\omega,I}, \quad x \in I, \end{aligned} \tag{5.1.46}$$

where

$$\omega = \left(\frac{\varphi^q}{\rho^q}\right)^{1/q-1} w^q. \tag{5.1.47}$$

Consequently, by Lemma 2.4.5,

$$\varphi(x) \approx \|\min\{\rho(\cdot), \rho(x)\}\|_{\infty,W_\infty,I},$$

where

$$W_\infty = \frac{\varphi}{\rho}. \tag{5.1.48}$$

Hence,

$$\varphi(x) \approx \|\min\{R(\cdot), R(x)\}\|_{\infty,W,I} =: \widetilde{\Phi}(x) \quad \text{for all } x \in I, \tag{5.1.49}$$

with

$$R = \rho \quad \text{and} \quad W = W_\infty. \tag{5.1.50}$$

Moreover, by (5.1.6) and (5.1.49),

$$\varphi(x) < +\infty \quad \text{and} \quad \widetilde{\Phi}(x) < +\infty \qquad \text{for all } x \in I.$$

Again, we are faced with a similar problem to that encountered in Case A-1-2, namely, that we cannot apply Theorem 4.2.3 directly as we don't know if $\{x_k\}_{k=K_-}^{K_+}$ is a covering sequence of the function $\widetilde{\Phi}$. The equivalence of conditions (5.1.27) and (5.1.39) in Case A-1-2 now has the analogue that (5.1.45) is equivalent to

$$\left\| \sup_{t \in \widetilde{I}_k} \widetilde{\Phi}(t)\, \|v^{-1}\|_{\infty,(t,\widetilde{x}_k)} \right\|_{\ell^\infty(\widetilde{\mathcal{K}}^+)} < +\infty, \tag{5.1.51}$$

where

$$\{\widetilde{x}_k\}_{k=\widetilde{K}_-}^{\widetilde{K}_+} \in CS(\widetilde{\Phi}, R, I, \beta) \quad \text{with } \beta > 1,$$

and $\widetilde{\mathcal{K}}^+$, $\widetilde{I}_k$ are given by (5.1.41), (5.1.42); this latter equivalence is proved similarly. An application of Theorem 4.2.3 (with $q = +\infty$) gives that (5.1.51) is equivalent to

$$\left\| \sup_{a<t\leq x} R(t) \, \|v^{-1}\|_{\infty,(t,b)} \right\|_{\infty,W,I} < +\infty, \tag{5.1.52}$$

and (5.1.2) follows on using (5.1.50) and (5.1.48) in (5.1.52). The proof of the theorem is therefore complete in Case A.

Case B: In this case the discrete condition (5.1.22) becomes

$$\left\| \sup_{t\in I_k} \varphi(t) \, \|v^{-1}\|_{p',(t,x_k)} \right\|_{\ell^r(\mathcal{K}^+)} < +\infty, \tag{5.1.53}$$

where $\frac{1}{r} = \frac{1}{q} - \frac{1}{p}$ and (cf. (5.1.7))

$$\{x_k\}_{k=K_-}^{K_+} \in CS(\varphi, \rho, I, \alpha) \quad \text{with } \alpha > 2^{1/q}.$$

It is necessary to consider separate cases within Case B.

Case B-1: $1 < p \leq +\infty$, $0 < q < p$. Under these circumstances, (5.1.53) is equivalent to

$$\left\| \sup_{t\in I_k} \varphi^{p'}(t) \int_t^{x_k} v^{-p'}(s) \, ds \right\|_{\ell^{r/p'}(\mathcal{K}^+)} < +\infty. \tag{5.1.54}$$

From (5.1.33), (5.1.34), and Lemma 2.4.5,

$$\Phi(x) := \varphi^{p'}(x) \approx \| \min\{R(\cdot), R(x)\}\|_{r/p',W,I} =: \widetilde{\Phi}(x) \quad \text{for all } x \in I, \tag{5.1.55}$$

where

$$R = \rho^{p'} \quad \text{and} \quad W = \left(\frac{\varphi}{\rho}\right)^{p'} \left(\left(\frac{\rho}{\varphi}\right)^{p'} \omega \right)^{p'/r} = \left(\frac{\varphi}{\rho}\right)^{p'} \left(\frac{\rho}{\varphi} w\right)^{qp'/r}. \tag{5.1.56}$$

Moreover, by (5.1.6) and (5.1.55),

$$\Phi(x) < +\infty \quad \text{and} \quad \widetilde{\Phi}(x) < +\infty \quad \text{for all } x \in I.$$

We claim that condition (5.1.54) is equivalent to

$$\left\| \sup_{t\in \widetilde{I}_k} \widetilde{\Phi}(t) \int_t^{\widetilde{x}_k} v^{-p'}(s) \, ds \right\|_{\ell^{r/p'}(\widetilde{\mathcal{K}}^+)} < +\infty, \tag{5.1.57}$$

where

$$\{\widetilde{x}_k\}_{k=\widetilde{K}_-}^{\widetilde{K}_+} \in CS(\widetilde{\Phi}, R, I, \beta) \quad \text{with } \beta > 2^{p'/r},$$

and $\widetilde{\mathcal{K}}^+$, $\widetilde{I}_k$ are given by (5.1.41), (5.1.42). An application of Theorem 4.2.1 (with $q=r/p'$) will then give that (5.1.57) is equivalent to

$$\left\|\sup_{a<t\le x} R(t)\int_t^b v^{-p'}(s)\,ds\right\|_{r/p',W,I}<+\infty.$$

As this condition can be rewritten as

$$\|W^{1/p'}(x)\sup_{a<t\le x} R^{1/p'}(t)\,\|v^{-1}\|_{p',(t,b)}\|_{r,I}<+\infty, \tag{5.1.58}$$

(5.1.3) will follow on using (5.1.56) and (5.1.4). The asserted equivalence of (5.1.54) and (5.1.57) is established analogously to the equivalence of (5.1.27) and (5.1.39) in Case A-1-2.

Case B-2: $p=1,\ 0<q<p$. Thus $p'=\infty$ and (5.1.53) becomes

$$\|\sup_{t\in I_k}\varphi(t)\,\|v^{-1}\|_{\infty,(t,x_k)}\|_{\ell^r(\mathcal{K}^+)}<+\infty. \tag{5.1.59}$$

By (5.1.4), we have (5.1.32), and, on applying Theorem 2.4.3, we arrive at

$$\begin{aligned}\varphi(x)=(\varphi^q(x))^{1/q}&\approx\textstyle\int_I\min\{\rho(t),\rho(x)\}\left(\frac{\varphi^q}{\rho^q}\right)^{1/q-1}(t)\,w^q(t)\,dt\\&=\|\min\{\rho(\cdot),\rho(x)\}\|_{1,\omega,I}\quad\text{for all } x\in I,\end{aligned}$$

where

$$\omega=\left(\frac{\varphi^q}{\rho^q}\right)^{1/q-1}w^q.$$

Lemma 2.4.5 then gives

$$\varphi(x)\approx\|\min\{\rho(\cdot),\rho(x)\}\|_{r,W_r,I}=:\widetilde{\Phi}(x)\quad\text{for all } x\in I, \tag{5.1.60}$$

with

$$W_r=\left(\frac{\varphi}{\rho}\right)\left(\frac{\rho\omega}{\varphi}\right)^{1/r}. \tag{5.1.61}$$

Moreover, by (5.1.6) and (5.1.60),

$$\varphi(x)<+\infty\quad\text{and}\quad\widetilde{\Phi}(x)<+\infty\quad\text{for all } x\in I.$$

As in the corresponding problems in previous cases, condition (5.1.59) can be shown to be equivalent to

$$\|\sup_{t\in\widetilde{I}_k}\widetilde{\Phi}(t)\,\|v^{-1}\|_{\infty,(t,\widetilde{x}_k)}\|_{\ell^r(\widetilde{\mathcal{K}}^+)}<+\infty, \tag{5.1.62}$$

where

$$\{\widetilde{x}_k\}_{k=\widetilde{K}_-}^{\widetilde{K}_+}\in CS(\widetilde{\Phi},R,I,\beta)\quad\text{with }\beta>2^{1/r},$$

and $\widetilde{\mathcal{K}}^+$, $\widetilde{I}_k$ are given by (5.1.41), (5.1.42). It follows by an application of Theorem 4.2.3 (with $q = r$) that condition (5.1.62) is equivalent to

$$\| \sup_{a<t\le x} \rho(t) \, \|v^{-1}\|_{\infty,(t,b)} \|_{r,W_r,I} < +\infty,$$

and this is readily verified to coincide with (5.1.3). The proof of Theorem 5.1.1 is therefore complete. □

5.2 The inequality $\|g\|_{p,v,I} \lesssim \left\| \sup_{a<t\le x} \rho(t) \int_t^b g(s)\, ds \right\|_{q,w,I}$

The next theorem deals with the case of $p \le 1$ and the reverse of the inequality (5.1.1) in Theorem 5.1.1.

Theorem 5.2.1 *Let $I = (a,b) \subseteq \mathbb{R}$, $\rho \in Ads(I), 0 < p \le 1$ and $0 < q < +\infty$. Let $w, v \in \mathcal{W}(I)$ and suppose that*

$$\|v\|_{p^*,(a,t)} < +\infty \quad \textit{for all} \quad t \in I, \tag{5.2.1}$$

where $p^* = p/(1-p)$. Then

$$\|g\|_{p,v,I} \lesssim \left\| \sup_{a<t\le x} \rho(t) \int_t^b g(s)\, ds \right\|_{q,w,I} \quad \textit{for all } g \in \mathcal{M}^+(I), \tag{5.2.2}$$

if and only if one of the following is satisfied:

Case A: $q \le p$ *and*

$$\begin{aligned} &\left\| \left[\int_I \min\left\{1, \frac{\rho^p(x)}{\rho^p(t)}\right\} d\|v\|^p_{p^*,(a,t)} \right]^{1/p} \| \min\{\rho(\cdot), \rho(x)\}\|^{-1}_{q,w,I} \right\|_{\infty,I} \\ &+ \frac{\lim_{x\to a+} \|v\|_{p^*,(a,x)}}{\lim_{x\to a+} \| \min\{\rho(\cdot),\rho(x)\}\|_{q,w,I}} < +\infty \end{aligned} \tag{5.2.3}$$

(note that the final term may be nonzero only if $p = 1$);

Case B: $p < q$ and

$$\begin{aligned} &\left(\int_I \left(\int_I \min\left\{1, \frac{\rho^p(x)}{\rho^p(t)}\right\} d\|v\|^p_{p^*,(a,t)} \right)^{\frac{r}{p}} \frac{\int_a^x \rho^q(s) w^q(s)\, ds \int_x^b w^q(s)\, ds}{\| \min\{\rho(\cdot),\rho(x)\}\|^{r+2q}_{q,w,I}} \, d\rho^q(x) \right)^{\frac{1}{r}} \\ &+ \frac{\left(\int_I \rho^{-p}(t)\, d\|v\|^p_{p^*,(a,t)}\right)^{\frac{1}{p}}}{\|w\|_{q,I}} + \frac{\left(\int_I d\|v\|^p_{p^*,(a,t)}\right)^{\frac{1}{p}}}{\|\rho w\|_{q,I}} \\ &+ \frac{\lim_{x\to a+} \|v\|_{p^*,(a,t)}}{\lim_{x\to a+} \| \min\{\rho(\cdot),\rho(x)\}\|_{q,w,I}} < +\infty, \end{aligned} \tag{5.2.4}$$

where $1/r := 1/p - 1/q$.

Remarks 5.2.2 (i) *The integral involved in* (5.2.3) is the Lebesgue-Stieltjes integral. Note also that Convention 1.1.1 (i) has to be used when one calculates the expression on the left-hand side of (5.2.3). In particular, this concerns the case when $\|\min\{\rho(\cdot),\rho(x)\}\|_{q,w,I} = +\infty$ for some $x \in I$. Recall that the function

$$\varphi(x) := \|\min\{\rho(\cdot),\rho(x)\}\|_{q,w,I}, \quad x \in I,$$

belongs to $Q_\rho(I)$. Consequently, $\varphi(x) = +\infty$ for some $x \in I$ if and only if $\varphi \equiv +\infty$.

Now let $p \in (0,1)$. Since then

$$\frac{d}{dt}\|v\|^p_{p^*,(a,t)} = (1-p)\Big(\int_a^t v^{p^*}(s)\,ds\Big)^{-p} v^{p^*}(t), \; t \in I, \tag{5.2.5}$$

we can rewrite (5.2.3) as

$$\Big\| \Big[\int_I \min\Big\{1, \frac{\rho^p(x)}{\rho^p(t)}\Big\} \Big(\int_a^t v^{p^*}(s)\,ds\Big)^{-p} v^{p^*}(t)\,dt\Big]^{1/p} \times \|\min\{\rho(\cdot),\rho(x)\}\|^{-1}_{q,w,I} \Big\|_{\infty,I} < +\infty. \tag{5.2.6}$$

(ii) If $p \in (0,1)$, then

$$\int_I d\|v\|^p_{p^*,(a,t)} = \|v\|^p_{p^*,I}\,.$$

Thus the term

$$\frac{\Big(\int_I d\|v\|^p_{p^*,(a,t)}\Big)^{1/p}}{\|\rho w\|_{q,I}}$$

appearing in (5.2.4) can be rewritten as

$$\frac{\|v\|_{p^*,I}}{\|\rho w\|_{q,I}}\,.$$

To prove Theorem 5.2.1 we shall need the following three lemmas. The first is the counterpart of Lemma 5.1.3.

Lemma 5.2.3 *Let* $I = (a,b) \subseteq \mathbb{R}$, $\rho \in Ads(I)$ *and* $w, v \in \mathcal{W}(I)$. *Assume that* $0 < p \le 1$ *and* $0 < q \le +\infty$. *Let*

$$\varphi(x) := \|\min\{\rho(\cdot),\rho(x)\}\|_{q,w,I} < +\infty \quad \text{for all } x \in I, \tag{5.2.7}$$

and

$$\{x_k\}_{k=K_-}^{K_+} \in CS(\varphi,\rho,I,\alpha) \quad \text{with} \quad \alpha > 2^{1/q}. \tag{5.2.8}$$

Define [3]

$$(5.2.9) \qquad B_k := \sup_{h \in \mathcal{M}^+(I_k)} \left(\|hv\|_{p,I_k} \Big/ \sup_{t \in I_k} \varphi(t) \int_t^{x_k} h(s)\, ds \right), \quad k \in \mathcal{K}^+,$$

where $I_k = (x_{k-1}, x_k]$. *Then inequality* (5.2.2) *holds if and only if*

$$(5.2.10) \qquad \|\{B_k\}\|_{\ell^r(\mathcal{K}^+)} < +\infty,$$

where

$$(5.2.11) \qquad \frac{1}{r} := \left(\frac{1}{p} - \frac{1}{q} \right)_+ .$$

Proof. As in the proof of Lemma 5.1.3, on using Theorem 4.2.1, we rewrite inequality (5.2.2) in the form

$$(5.2.12) \quad \| \, \|gv\|_{p,I_k} \|_{\ell^p(\mathcal{K}^+)} \lesssim \left\| \sup_{t \in I_k} \varphi(t) \int_t^{x_k} g(s)\, ds \right\|_{\ell^q(\mathcal{K}^+)} \quad \text{for all } g \in \mathcal{M}^+(I).$$

To find a sufficient condition for the validity of inequality (5.2.12), we apply locally (that is, for any $k \in \mathcal{K}^+$) the reverse Hardy-type inequality

$$(5.2.13) \qquad \|hv\|_{p,I_k} \le B_k \sup_{t \in I_k} \varphi(t) \int_t^{x_k} h(s)\, ds, \quad h \in \mathcal{M}^+(I_k),$$

where the best possible constant B_k is that defined in (5.2.9).

If $0 < q \le p$, then using (5.2.13)) and the Jensen inequality (cf. [HLP, Theorem 19, p. 28])

$$\left(\sum_{k \in \mathcal{K}^+} a_k^p \right)^{1/p} \le \left(\sum_{k \in \mathcal{K}^+} a_k^q \right)^{1/q}, \quad 0 < q \le p,$$

(which holds for all sequences $\{a_k\}_{k \in \mathcal{K}^+}$ of non-negative numbers), we obtain

$$\begin{aligned} \text{LHS}(5.2.12) &\le \left\| B_k \sup_{t \in I_k} \varphi(t) \int_t^{x_k} g(s)\, ds \right\|_{\ell^p(\mathcal{K}^+)} \\ &\le \|\{B_k\}\|_{\ell^\infty(\mathcal{K}^+)} \left\| \sup_{t \in I_k} \varphi(t) \int_t^{x_k} g(s)\, ds \right\|_{\ell^p(\mathcal{K}^+)} \\ &\le \|\{B_k\}\|_{\ell^\infty(\mathcal{K}^+)} \, \text{RHS}(5.2.12). \end{aligned}$$

If $p < q = +\infty$, then $r = p$ and thus, by (5.2.13),

$$\begin{aligned} \text{LHS}(5.2.12) &\le \left\| B_k \sup_{t \in I_k} \varphi(t) \int_t^{x_k} g(s)\, ds \right\|_{\ell^p(\mathcal{K}^+)} \\ &\le \|\{B_k\}\|_{\ell^p(\mathcal{K}^+)} \left\| \sup_{t \in I_k} \varphi(t) \int_t^{x_k} g(s)\, ds \right\|_{\ell^\infty(\mathcal{K}^+)} \\ &= \|\{B_k\}\|_{\ell^r(\mathcal{K}^+)} \, \text{RHS}(5.2.12). \end{aligned}$$

[3] Note that in (5.2.9) we use the convention that $0/0 = 0$ and $(+\infty)/(+\infty) = 0$ - cf. Convention 1.1.1 (i).

In the case that $p < q < +\infty$, Hölder's inequality for sums (with the exponents r/p and q/p) applied to (5.2.13) gives

$$\begin{aligned}
[\mathrm{LHS}(5.2.12)]^p &= \sum_{k\in\mathcal{K}^+} \|gv\|_{p,I_k}^p \\
&\le \sum_{k\in\mathcal{K}^+} \Big[B_k \sup_{t\in I_k} \varphi(t) \int_t^{x_k} g(s)\,ds \Big]^p \\
&\le \Big(\sum_{k\in\mathcal{K}^+} B_k^r \Big)^{p/r} \Big(\sum_{k\in\mathcal{K}^+} \Big[\sup_{t\in I_k} \varphi(t) \int_t^{x_k} g(s)\,ds \Big]^q \Big)^{p/q} \\
&= \|\{B_k\}\|_{\ell^r(\mathcal{K}^+)}^p \, [\mathrm{RHS}(5.2.12)]^p.
\end{aligned}$$

Therefore in all cases, (5.2.12), and hence (5.2.2), holds if condition (5.2.10) is satisfied. (Note that inequality (5.2.13) holds with a finite constant B_k, $k \in \mathcal{K}^+$, if (5.2.10) is fulfilled.)

The proof that condition (5.2.10) is also necessary for the validity of inequality (5.2.12) (and so for (5.2.2), too) is analogous to the corresponding part of that of Lemma 5.1.3. □

Lemma 5.2.4 *Let $I = (a,b) \subseteq \mathbb{R}$, $\rho \in Ads(I)$, $0 < p \le 1$, $0 < q \le +\infty$ and $w, v \in \mathcal{W}(I)$. Suppose that (5.2.1), (5.2.7), (5.2.8) are satisfied and that $B_k, k \in \mathcal{K}^+$, are given by (5.2.9). Define $\psi(t) := \|v\|_{p^*,(a,t)}^p$ for all $t \in I$, and*

$$\widehat{B}_k := \Big(\int_{(x_{k-1},x_k)} \varphi^{-p}(t)\,d\psi(t) \Big)^{1/p}, \quad k \in \mathcal{K}^+. \tag{5.2.14}$$

Then

$$B_k^p \lesssim \widehat{B}_k^p + \frac{\psi(x_{k-1}+)}{\varphi^p(x_{k-1})}, \tag{5.2.15}$$

$$\widehat{B}_k^p \lesssim B_k^p, \tag{5.2.16}$$

for all $k \in \mathcal{K}^+$.

Proof. By Theorem 1.5.1, for all $k \in \mathcal{K}^+$,

$$B_k^p \approx \int_{(x_{k-1},x_k)} \|v\|_{p^*,(x_{k-1},t]}^p \, d[-\varphi^{-p}(t)] + \frac{\|v\|_{p^*,(x_{k-1},x_k)}^p}{\varphi^p(x_k)}. \tag{5.2.17}$$

Moreover, since

$$\|v\|_{p^*,(x_{k-1},t]}^p = \|v\|_{p^*,(x_{k-1},t)}^p \quad \text{for all } t \in (x_{k-1},x_k),$$

by Theorem 1.5.2, for all $k \in \mathcal{K}^+$,

$$B_k^p \approx \int_{(x_{k-1},x_k)} \varphi^{-p}(t)\, d[\,\|v\|_{p^*,(x_{k-1},t+]}^p\,] + \frac{\|v\|_{p^*,(x_{k-1},x_{k-1}+)}^p}{\varphi^p(x_{k-1})}. \tag{5.2.18}$$

Thus, to prove (5.2.16), it is sufficient to verify that

$$\int_{(x_{k-1},x_k)} \varphi^{-p}(t)\, d\psi(t) \le \int_{(x_{k-1},x_k)} \varphi^{-p}(t)\, d[\, \|v\|^p_{p^*,(x_{k-1},t+]}\,] + \frac{\|v\|^p_{p^*,(x_{k-1},x_{k-1}+)}}{\varphi^p(x_{k-1})}. \tag{5.2.19}$$

Since $\varphi^{-p} \in \mathcal{M}^+(I;\downarrow)$, a simple argument using the monotone convergence theorem shows that (5.2.19) holds provided that, for all $y \in (x_{k-1}, x_k)$,

$$\int_{(x_{k-1},y)} d\psi(t) \le \int_{(x_{k-1},y)} d[\, \|v\|^p_{p^*,(x_{k-1},t+]}\,] + \|v\|^p_{p^*,(x_{k-1},x_{k-1}+)}.$$

The last inequality is satisfied since it can be rewritten as

$$\|v\|^p_{p^*,(a,y)} \le \|v\|^p_{p^*,(a,x_{k-1}+)} + \|v\|^p_{p^*,(x_{k-1},y)}.$$

Therefore, (5.2.16) is proved.

To prove (5.2.15), it is sufficient to verify that

$$\text{RHS}(5.2.18) \lesssim \text{RHS}(5.2.15).$$

Hence, using the same argument as that in the proof of inequality (5.2.19), we see that it is enough to show that, for all $y \in (x_{k-1}, x_k)$,

$$\int_{(x_{k-1},y)} d[\, \|v\|^p_{p^*,(x_{k-1},t+]}\,] + \|v\|^p_{p^*,(x_{k-1},x_{k-1}+)} \lesssim \int_{(x_{k-1},y)} d[\, \|v\|^p_{p^*,(a,t)}\,] + \|v\|^p_{p^*,(a,x_{k-1}+)}.$$

This estimate is satisfied since it can be rewritten as

$$\|v\|^p_{p^*,(x_{k-1},y)} \lesssim \|v\|^p_{p^*,(a,y)}.$$

Hence, (5.2.15) is proved. □

Corollary 5.2.5 *Let all the assumptions of* Lemma 5.2.4 *be satisfied. Define*

$$\tilde{B}_k := \Big(\int_{(x_{k-1},x_k]\cap I} \varphi^{-p}(t)\, d\psi(t) \Big)^{1/p}, \quad k \in \mathcal{K}^+. \tag{5.2.20}$$

Then

$$B^p_k \lesssim \tilde{B}^p_k + \frac{\psi(x_{k-1}+)}{\varphi^p(x_{k-1})}, \quad \textit{for all } k \in \mathcal{K}^+, \tag{5.2.21}$$

$$\tilde{B}^p_k \lesssim B^p_k + \frac{\psi(x_k+)}{\varphi^p(x_k)}, \quad \textit{for all } k \in \mathcal{K}, \tag{5.2.22}$$

and, if $K_+ < +\infty$, *then*

$$\tilde{B}^p_{K_+} \lesssim B^p_{K_+}. \tag{5.2.23}$$

Proof. Since $(x_{k-1}, x_k) \subset (x_{k-1}, x_k] \cap I$, we see that $\widehat{B}_k \le \tilde{B}_k$ for all $k \in \mathcal{K}^+$, with $\widehat{B}_k$ defined in (5.2.14). Thus (5.2.21) is a consequence of (5.2.15).

If $k \in \mathcal{K}$, we obtain

$$
\begin{aligned}
\tilde{B}_k^p &= \int_{(x_{k-1},x_k]} \varphi^{-p}(t)\,d\psi(t) \\
&= \int_{(x_{k-1},x_k)} \varphi^{-p}(t)\,d\psi(t) + \frac{\psi(x_k+) - \psi(x_k-)}{\varphi^p(x_k)} \\
&\le \int_{(x_{k-1},x_k)} \varphi^{-p}(t)\,d\psi(t) + \frac{\psi(x_k+)}{\varphi^p(x_k)} \\
&= \widehat{B}_k^p + \frac{\psi(x_k+)}{\varphi^p(x_k)} \\
&\le B_k^p + \frac{\psi(x_k+)}{\varphi^p(x_k)}
\end{aligned}
$$

on using (5.2.16), and (5.2.22) follows.

Finally, if $k = K_+ < +\infty$, then $(x_{k-1}, x_k] \cap I = (x_{k-1}, x_k)$ and so $\tilde{B}_k = \widehat{B}_k$. Therefore, (5.2.23) is a consequence of (5.2.16).

□

Lemma 5.2.6 *Let all the assumptions of* Lemma 5.2.4 *be satisfied and let* $\{\tilde{B}_k\}$, $k \in \mathcal{K}^+$, *be given by* (5.2.20). *Then*

$$
\|\{B_k\}\|_{\ell^r(\mathcal{K}^+)} \approx \|\{\tilde{B}_k\}\|_{\ell^r(\mathcal{K}^+)} + \frac{\psi^{1/p}(a)}{\varphi(a)}, \quad 0 < r \le +\infty, \tag{5.2.24}
$$

(note that the final term may be nonzero only if $p = 1$; recall also that $\psi(a) := \psi(a+), \varphi(a) := \varphi(a+)$).

Proof. By Corollary 5.2.5,

$$
\|\{B_k\}\|_{\ell^r(\mathcal{K}^+)} \lesssim \left\|\left\{\tilde{B}_k\right\}\right\|_{\ell^r(\mathcal{K}^+)} + \left\|\left\{\frac{\psi^{1/p}(x_{k-1}+)}{\varphi(x_{k-1})}\right\}\right\|_{\ell^r(\mathcal{K}^+)}. \tag{5.2.25}
$$

If we prove that

$$
\left\|\left\{\frac{\psi^{1/p}(x_{k-1}+)}{\varphi(x_{k-1})}\right\}\right\|_{\ell^r(\mathcal{K}^+)} \lesssim \|\{\tilde{B}_k\}\|_{\ell^r(\mathcal{K}^+)} + \frac{\psi^{1/p}(a)}{\varphi(a)}, \tag{5.2.26}
$$

then (5.2.25) and (5.2.26) will imply that

$$
\|\{B_k\}\|_{\ell^r(\mathcal{K}^+)} \lesssim \|\{\tilde{B}_k\}\|_{\ell^r(\mathcal{K}^+)} + \frac{\psi^{1/p}(a)}{\varphi(a)}, \tag{5.2.27}
$$

which means that LHS((5.2.24))$\lesssim$ RHS((5.2.24)).

We now verify estimate (5.2.26). If $J_- > -\infty$, $k \in \mathcal{K}^+$, and $k \geq J_- + 1$, then since $K_- = J_- - 1$,

$$\psi(x_{k-1}+) = \sum_{i=J_-+1}^{k} [\psi(x_{i-1}+) - \psi(x_{i-2}+)] + \psi(x_{K_-}+) \tag{5.2.28}$$

$$= \sum_{i=J_-+1}^{k} [\psi(x_{i-1}+) - \psi(x_{i-2}+)] + \psi(a).$$

If $J_- = -\infty$, $k, j \in \mathcal{K}^+$, and $k \geq j$, then

$$\psi(x_{k-1}+) = \sum_{i=j}^{k} [\psi(x_{i-1}+) - \psi(x_{i-2}+)] + \psi(x_{j-2}+),$$

which, on letting $j \to -\infty$, yields

$$\psi(x_{k-1}+) = \sum_{i=-\infty}^{k} [\psi(x_{i-1}+) - \psi(x_{i-2}+)] + \psi(a+) \tag{5.2.29}$$

$$= \sum_{i=J_-+1}^{k} [\psi(x_{i-1}+) - \psi(x_{i-2}+)] + \psi(a).$$

Estimates (5.2.28) and (5.2.29) show that if $k \in \mathcal{K}^+$ and $k \geq J_- + 1$, then

$$\psi(x_{k-1}+) = \sum_{i=J_-+1}^{k} [\psi(x_{i-1}+) - \psi(x_{i-2}+)] + \psi(a), \tag{5.2.30}$$

which can be rewritten as

$$\psi(x_{k-1}+) = \sum_{i=J_-+1}^{k} \int_{(x_{i-2}, x_{i-1}]} d\psi(t) + \psi(a). \tag{5.2.31}$$

If $m_1, m_2 \in \overline{\mathbb{Z}}$, $m_1 < m_2$, we put

$$S(m_1, m_2) := \{k \in \mathbb{Z}; m_1 \leq k \leq m_2\}.$$

On using (5.2.31), the fact that the sequence $\{\varphi^{-p}(x_k)\}$ is geometrically decreasing, Lemmas 1.3.3 and 1.3.4, we obtain

$$
\begin{aligned}
(5.2.32)\quad & \left\| \left\{ \frac{\psi^{1/p}(x_{k-1}+)}{\varphi(x_{k-1})} \right\} \right\|_{\ell^r(S(J_-+1,K_+))} \\
&\lesssim \left\| \frac{1}{\varphi(x_{k-1})} \left(\sum_{i=J_-+1}^{k} \int_{(x_{i-2},x_{i-1}]} d\psi(t) \right)^{1/p} \right\|_{\ell^r(S(J_-+1,K_+))} \\
&\quad + \psi^{1/p}(a) \left\| \frac{1}{\varphi(x_{k-1})} \right\|_{\ell^r(S(J_-+1,K_+))} \\
&\approx \left\| \frac{1}{\varphi^p(x_{k-1})} \sum_{i=J_-+1}^{k} \int_{(x_{i-2},x_{i-1}]} d\psi(t) \right\|_{\ell^{r/p}(S(J_-+1,K_+))}^{1/p} + \frac{\psi^{1/p}(a)}{\varphi(a)} \\
&\approx \left\| \frac{1}{\varphi^p(x_{k-1})} \int_{(x_{k-2},x_{k-1}]} d\psi(t) \right\|_{\ell^{r/p}(S(J_-+1,K_+))}^{1/p} + \frac{\psi^{1/p}(a)}{\varphi(a)} \\
&\le \left\| \left(\int_{(x_{k-2},x_{k-1}]} \varphi^{-p}(t)\, d\psi(t) \right)^{1/p} \right\|_{\ell^r(S(J_-+1,K_+))} + \frac{\psi^{1/p}(a)}{\varphi(a)} \\
&= \|\{\tilde{B}_{k-1}\}\|_{\ell^r(S(J_-+1,K_+))} + \frac{\psi^{1/p}(a)}{\varphi(a)} \\
&= \|\{\tilde{B}_{k}\}\|_{\ell^r(S(J_-,J_+))} + \frac{\psi^{1/p}(a)}{\varphi(a)} \\
&\le \|\{\tilde{B}_{k}\}\|_{\ell^r(\mathcal{K}^+)} + \frac{\psi^{1/p}(a)}{\varphi(a)}.
\end{aligned}
$$

If $J_- = -\infty$, then $S(J_-+1, K_+) = \mathcal{K}^+$ and (5.2.32) implies (5.2.26). If $J_- > -\infty$, then

$$\mathcal{K}^+ = S(J_-+1, K_+) \cup \{J_-\}.$$

Moreover, if $k = J_-$, then

$$\frac{\psi^{1/p}(x_{k-1}+)}{\varphi(x_{k-1})} = \frac{\psi^{1/p}(a)}{\varphi(a)}.$$

Together with estimate (5.2.32), this yields

$$
\begin{aligned}
\left\| \left\{ \frac{\psi^{1/p}(x_{k-1}+)}{\varphi(x_{k-1})} \right\} \right\|_{\ell^r(\mathcal{K}^+))} &\approx \left\| \left\{ \frac{\psi^{1/p}(x_{k-1}+)}{\varphi(x_{k-1})} \right\} \right\|_{\ell^r(S(J_-+1,K_+))} + \frac{\psi^{1/p}(a)}{\varphi(a)} \\
&\lesssim \|\{\tilde{B}_k\}\|_{\ell^r(\mathcal{K}^+)} + \frac{\psi^{1/p}(a)}{\varphi(a)},
\end{aligned}
$$

and (5.2.26) follows. Thus (5.2.27) is proved.

Next, we prove the reverse estimate to (5.2.27), i.e.,

$$\|\{B_k\}\|_{\ell^r(\mathcal{K}^+)} \gtrsim \|\{\tilde{B}_k\}\|_{\ell^r(\mathcal{K}^+)} + \frac{\psi^{1/p}(a)}{\varphi(a)}. \tag{5.2.33}$$

By Corollary 5.2.5,

$$\|\{\tilde{B}_k\}\|_{\ell^r(\mathcal{K}^+)} \lesssim \|\{B_k\}\|_{\ell^r(\mathcal{K}^+)} + \left\|\left\{\frac{\psi^{1/p}(x_k+)}{\varphi(x_k)}\right\}\right\|_{\ell^r(\mathcal{K})}. \tag{5.2.34}$$

Put

$$\psi_k(t) := \|v\|^p_{p*,(x_{k-1},t)} \quad \text{if } k \in \mathcal{K} \text{ and } t \in (x_{k-1}, x_{k+1}).$$

If $0 < p < 1$, then, for $t \in (x_{k-1}, x_k]$, $k \in \mathcal{K}$,

$$\begin{aligned} \psi(t+) &= \left(\int_{(a,x_{k-1})} |v(x)|^{p^*}\,dx + \int_{(x_{k-1},t+)} |v(x)|^{p^*}\,dx\right)^{1-p} \\ &= \left(\psi(x_{k-1})^{1/(1-p)} + \psi_k(t+)^{1/(1-p)}\right)^{1-p} \\ &\le \psi(x_{k-1}) + \psi_k(t+) \\ &\le \psi(x_{k-1}+) + \psi_k(t+), \end{aligned} \tag{5.2.35}$$

and for $p = 1$,

$$\psi(t+) \le \psi(x_{k-1}) + \psi_k(t+) \le \psi(x_{k-1}+) + \psi_k(t+). \tag{5.2.36}$$

From (5.2.35) and (5.2.36), we infer that

$$\psi(x_k+) - \psi(x_{k-1}+) \le \psi_k(x_k+) \quad \text{for all } k \in \mathcal{K}.$$

Consequently,

$$\begin{aligned} \psi(x_k+) &= \sum_{j=J_-}^{k} [\psi(x_j+) - \psi(x_{j-1}+)] + \psi(a) \\ &\le \sum_{j=J_-}^{k} \psi_j(x_j+) + \psi(a) \quad \text{for all } k \in \mathcal{K}. \end{aligned} \tag{5.2.37}$$

Since the sequence $\{\varphi^{-p}(x_k)\}$ is geometrically decreasing, on applying (5.2.37), Lemmas 1.3.3 and Lemma 1.3.4, we arrive at

$$\left\| \left\{ \frac{\psi^{1/p}(x_k+)}{\varphi(x_k)} \right\} \right\|_{\ell^r(\mathcal{K})} \tag{5.2.38}$$
$$\lesssim \left\| \left\{ \frac{1}{\varphi(x_k)} \left(\sum_{j=J_-}^{k} \psi_j(x_j+) \right)^{1/p} \right\} \right\|_{\ell^r(\mathcal{K})} + \frac{\psi^{1/p}(a)}{\varphi(a)}$$
$$= \left\| \left\{ \frac{1}{\varphi^p(x_k)} \left(\sum_{j=J_-}^{k} \psi_j(x_j+) \right) \right\} \right\|_{\ell^{r/p}(\mathcal{K})}^{1/p} + \frac{\psi^{1/p}(a)}{\varphi(a)}$$
$$\approx \left\| \left\{ \frac{\psi_k^{1/p}(x_k+)}{\varphi(x_k)} \right\} \right\|_{\ell^r(\mathcal{K})} + \frac{\psi^{1/p}(a)}{\varphi(a)}.$$

Therefore, to verify (5.2.33), it is sufficient to prove that

$$\left\| \left\{ \frac{\psi_k^{1/p}(x_k+)}{\varphi(x_k)} \right\} \right\|_{\ell^r(\mathcal{K})} \lesssim \|\{B_k\}\|_{\ell^r(\mathcal{K}^+)} \tag{5.2.39}$$

and

$$\frac{\psi^{1/p}(a)}{\varphi(a)} \lesssim \|\{B_k\}\|_{\ell^r(\mathcal{K}^+)}. \tag{5.2.40}$$

If $\psi_k(x_k+) = \psi_k(x_k)$ for some $k \in \mathcal{K}$, then, by (5.2.17),

$$\frac{\psi_k^{1/p}(x_k+)}{\varphi(x_k)} = \frac{\psi_k^{1/p}(x_k)}{\varphi(x_k)} \lesssim B_k; \tag{5.2.41}$$

in particular, (5.2.41) holds for all $k \in \mathcal{K}$ if $0 < p < 1$ since ψ_k is then continuous.

Suppose now that $p = 1$ and $k \in \mathcal{K}$ is such that

$$\psi_k(x_k+) > \psi_k(x_k). \tag{5.2.42}$$

If $k \in [J_-, J_+)$, then (5.2.8) implies that $\varphi(x_{k+1}) \ge \alpha\varphi(x_k)$, with $\alpha > 1$. Hence,

$$\frac{1}{\varphi(x_k)} \le \frac{\alpha}{\alpha - 1} \left(\frac{1}{\varphi(x_k)} - \frac{1}{\varphi(x_{k+1})} \right). \tag{5.2.43}$$

Thus, on using (5.2.43) and (5.2.17), we conclude that

$$\frac{\psi_k(x_k+)}{\varphi(x_k)} \lesssim \psi_k(x_k+) \int_{(x_k,x_{k+1})} d[-\varphi(t)^{-1}] \tag{5.2.44}$$
$$\le \int_{(x_k,x_{k+1})} \psi_{k+1}(t)\, d[-\varphi(t)^{-1}]$$
$$\lesssim B_{k+1}.$$

If $k = J_+$ (hence $J_+ < +\infty$), then

$$\psi_k(x_k+) = \|v\|_{\infty,(x_{k-1},x_k+)} \le \|v\|_{\infty,(x_{k-1},x_k)} + \|v\|_{\infty,(x_k,x_k+)}.$$

Consequently,

$$\frac{\psi_k(x_k+)}{\varphi(x_k)} \le \frac{\|v\|_{\infty,(x_{k-1},x_k)}}{\varphi(x_k)} + \frac{\|v\|_{\infty,(x_k,x_k+)}}{\varphi(x_k)},$$

and, on using (5.2.17) and (5.2.18), we obtain

$$\frac{\psi_k(x_k+)}{\varphi(x_k)} \lesssim B_k + B_{k+1}. \tag{5.2.45}$$

Estimates (5.2.41), (5.2.44) and (5.2.45) imply that

$$\left\| \left\{ \frac{\psi_k(x_k+)}{\varphi(x_k)} \right\} \right\|_{\ell^r(\mathcal{K})} \lesssim \|\{B_k\}\|_{\ell^r(\mathcal{K}^+)},$$

and so (5.2.39) is proved.

To verify estimate (5.2.40), we distinguish two cases.

Assume first that $k = K_- > -\infty$. Then

$$\psi(a) = \psi(a+) = \psi(x_k+) = \psi_{J_-}(x_k+).$$

Moreover,

$$\psi_{J_-}(x_k+) \le \|v\|^p_{p^*,(a,x_{J_-})}.$$

This estimate and (5.2.18) yield

$$\frac{\psi(a)}{\varphi^p(a)} \le \frac{\|v\|^p_{p^*,(a,x_{J_-})}}{\varphi^p(a)} = \frac{\|v\|^p_{p^*,(x_{J_--1},x_{J_-})}}{\varphi^p(x_{J_--1})} \lesssim B^p_{J_-}.$$

Assume now that $K_- = -\infty$. If

$$\psi(a) = \lim_{t\to a+} \|v\|^p_{p^*,(a,t)} = 0, \tag{5.2.46}$$

then (5.2.40) is trivial. In particular, (5.2.40) holds if $0 < p < 1$. Thus, it remains to prove (5.2.40) provided that $\psi(a) > 0$, $p = 1$ and $K_- = -\infty$. Then (cf. (5.2.17)), for all $k \in \mathcal{K}^+$,

$$\begin{aligned} \frac{\psi(x_k)}{\varphi(x_k)} &= \sup_{-\infty<j\le k} \frac{\|v\|_{\infty,(x_{j-1},x_j)}}{\varphi(x_k)} \\ &\le \sup_{-\infty<j\le k} \frac{\|v\|_{\infty,(x_{j-1},x_j)}}{\varphi(x_j)} \\ &\lesssim B_j \\ &\le \|\{B_j\}\|_{\ell^\infty(\mathcal{K}^+)}. \end{aligned} \tag{5.2.47}$$

Moreover, since $\ell^r(\mathcal{K}^+) \hookrightarrow \ell^\infty(\mathcal{K}^+)$ for any $r \in (0,+\infty]$, we obtain from (5.2.47) that

$$\frac{\psi(x_k)}{\varphi(x_k)} \lesssim \|\{B_j\}\|_{\ell^r(\mathcal{K}^+)} \quad \text{for all } k \in \mathcal{K}^+.$$

Consequently,

$$\frac{\psi(a)}{\varphi(a)} = \frac{\lim_{k\to-\infty} \psi(x_k)}{\lim_{k\to-\infty} \varphi(x_k)} = \lim_{k\to-\infty} \frac{\psi(x_k)}{\varphi(x_k)} \lesssim \|\{B_j\}\|_{\ell^r(\mathcal{K}^+)},$$

which means that (5.2.40) again holds. □

Proof of Theorem 5.2.1. Assume first that [4]

$$\varphi(x) := \|\min\{\rho(\cdot), \rho(x)\}\|_{q,\omega,I} = +\infty \quad \text{for some} \quad x \in I. \tag{5.2.48}$$

Let f be given by (5.1.24), i.e.,

$$f(x) = \sup_{a<t\le x} \rho(t) \int_t^b g(s)ds, \quad g \in \mathcal{M}^+(I).$$

Then $f \in Q_\rho(I)$ and, by Remark 4.1.2, $\|f\|_{q,w,I} = +\infty$ if $g \ne 0$ a.e. in I. Thus, RHS(5.2.2) $= +\infty$ for any $g \in \mathcal{M}^+(I)$ such that $g \ne 0$ a.e. in I. Consequently, (5.2.2) holds. On the other hand, by Convention 1.1.1 (i), (5.2.3) is satisfied as well. We claim that (5.2.4) is also satisfied. The finiteness of the first and last terms in (5.2.4) follows by our Convention 1.1.1 (i). Furthermore, since

$$\min\{\rho(t), \rho(x)\} \le \rho(x) \quad \text{and} \quad \min\{\rho(t), \rho(x)\} \le \rho(t) \qquad \text{for all } t, x \in I,$$

we obtain the estimates

$$\frac{\varphi(x)}{\rho(x)} \le \|w\|_{q,I} \quad \text{and} \quad \varphi(x) \le \|\rho w\|_{q,I} \qquad \text{for any } x \in I,$$

which imply that the middle two terms in (5.2.4) are zero. Hence the condition (5.2.4) is satisfied.

Now suppose that

$$\varphi(x) < +\infty \quad \text{for all} \quad x \in I, \tag{5.2.49}$$

and let

$$\{x_k\}_{k=K_-}^{K_+} \in CS(\varphi, \rho, I, \alpha) \quad \text{with} \quad \alpha > 2^{1/q}. \tag{5.2.50}$$

Then from Lemma 5.2.3 and Lemma 5.2.6 (with r given in (5.2.11)), the inequality (5.2.2) is equivalent to the discrete condition

$$\|\{\tilde{B}_k\}\|_{\ell^r(\mathcal{K}^+)} + \frac{\psi^{1/p}(a)}{\varphi(a)} < +\infty. \tag{5.2.51}$$

---

[4] Note that (5.2.48) implies that $\varphi \equiv +\infty$ on I.

It remains to prove that (5.2.51) with $r = \infty$, is equivalent to the "continuous criterion" (5.2.3) in Case A, and that (5.2.51) with $1/r = 1/p - 1/q$, is equivalent to (5.2.4) in Case B.

Put

$$B := \|\{\tilde{B}_k\}\|_{\ell^r(\mathcal{K}^+)}. \tag{5.2.52}$$

Then (5.2.51) can be rewritten as

$$B + \frac{\psi^{1/p}(a)}{\varphi(a)} < +\infty. \tag{5.2.53}$$

Case A: $0 < q \le p \le 1$. In this case, $r = +\infty$. From (5.2.52) and (5.2.20) we obtain

$$\begin{aligned} B^p &= \|\{\tilde{B}_k^p\}\|_{\ell^\infty(\mathcal{K}^+)} \\ &= \left\| \int_{I_k \cap I} \left(\frac{\rho}{\varphi}\right)^p(t) \, \frac{d\psi(t)}{\rho^p(t)} \right\|_{\ell^\infty(\mathcal{K}^+)} \\ &= \left\| \int_{I_k} \left(\frac{\rho}{\varphi}\right)^p(t) \, d\bar{\nu} \right\|_{\ell^\infty(\mathcal{K}^+)}, \end{aligned} \tag{5.2.54}$$

where $I_k := (x_{k-1}, x_k]$, $k \in \mathcal{K}^+$, ν is the Borel measure on I given by

$$d\nu := \rho^{-p}(t) \, d\psi(t) \tag{5.2.55}$$

and $\bar{\nu}$ is its extension by zero in $\mathbb{R} \setminus I$ (cf. page 2). The definition of φ in (5.2.7) implies that

$$\varphi^q(x) = \int_I \min\{\rho^q(t), \rho^q(x)\} \, w^q(t) \, dt =: H(x), \quad x \in I. \tag{5.2.56}$$

Thus, by Lemma 2.4.6,

$$\left(\frac{\rho}{\varphi}\right)^q(x) \approx \|\min\{\rho^q(\cdot), \rho^q(x)\}\|_{\infty, 1/H, I}, \quad x \in I,$$

which immediately yields

$$\Phi(x) := \left(\frac{\rho}{\varphi}\right)^p(x) = \left[\left(\frac{\rho}{\varphi}\right)^q(x)\right]^{p/q} \approx \widetilde{\Phi}(x), \quad x \in I,$$

where

$$\widetilde{\Phi}(x) := \|\min\{\rho^p(\cdot), \rho^p(x)\}\|_{\infty, H^{-p/q}, I} \quad x \in I.$$

Consequently,

$$B^p = \left\| \int_{I_k} \Phi(t) \, d\bar{\nu} \right\|_{\ell^\infty(\mathcal{K}^+)} \tag{5.2.57}$$

and $\widetilde{\Phi}$ is the ρ^p-fundamental function of the space $L^\infty(H^{-p/q}, I)$. By Remark 4.2.6 (with $q = \infty$),

$$(5.2.58) \qquad \left\| \int_{I_k} \Phi(t)\, d\bar{\nu} \right\|_{\ell^\infty(\mathcal{K}^+)} \approx \left\| \frac{\Phi(x_k)}{\rho^p(x_k)} \| \min\{\rho^p(\cdot), \rho^p(x_k)\}\|_{1,I,\nu} \right\|_{\ell^\infty(\mathcal{K}^+_-)}$$

and, by (5.2.50),

$$\{x_k\}_{k=K_-}^{K_+} \in CS(\Phi, \rho^p, I, \alpha^p) \quad \text{and} \quad \alpha^p > 1.$$

As $\Phi \approx \widetilde{\Phi}$ on I, on using Corollary 4.2.10, we see that

$$(5.2.59) \qquad \mathrm{RHS}(5.2.58) \approx \left\| \frac{\widetilde{\Phi}(\widetilde{x}_k)}{\rho^p(\widetilde{x}_k)} \| \min\{\rho^p(\cdot), \rho^p(\widetilde{x}_k)\}\|_{1,I,\nu} \right\|_{\ell^\infty(\widetilde{\mathcal{K}}^+_-)},$$

where

$$\{\widetilde{x}_k\}_{k=\widetilde{K}_-}^{\widetilde{K}_+} \in CS(\widetilde{\Phi}, \rho^p, I, \beta) \quad \text{with } \beta > 1$$

and

$$\widetilde{\mathcal{K}}^+_- := \{k \in \mathbb{Z}; \widetilde{K}_- \le k \le \widetilde{K}_+\}.$$

Now, by Theorem 4.2.5 (with $q = \infty$),

$$\mathrm{RHS}(5.2.59) \approx \| \, \| \min\{\rho^p(\cdot), \rho^p(x)\}\|_{1,I,\nu} \|_{\infty, H^{-p/q}, I}\,.$$

Consequently (cf. (5.2.57)-(5.2.59)),

$$B^p \approx \| \, \| \min\{\rho^p(\cdot), \rho^p(x)\}\|_{1,I,\nu} \|_{\infty, H^{-p/q}, I}.$$

On using this estimate, one can readily verify that (5.2.53) is equivalent to (5.2.3).

Case B: $0 < p < q < +\infty$ and $0 < p \le 1$. In this case $1/r = 1/p - 1/q$. From (5.2.52) and (5.2.20) we obtain

$$(5.2.60) \qquad \begin{aligned} B^p &= \|\{\tilde{B}_k\}\|^p_{\ell^r(\mathcal{K}^+)} \\ &= \left\| \int_{I_k \cap I} \left(\frac{\rho}{\varphi}\right)^p (t) \frac{d\psi(t)}{\rho^p(t)} \right\|_{\ell^{r/p}(\mathcal{K}^+)} \\ &= \left\| \int_{I_k} \left(\frac{\rho}{\varphi}\right)^p (t)\, d\bar{\nu} \right\|_{\ell^{r/p}(\mathcal{K}^+)}, \end{aligned}$$

where ν is the Borel measure on I given by (5.2.55) and $\bar{\nu}$ is its extension by zero in $\mathbb{R} \setminus I$.

Putting

$$(5.2.61) \qquad \Phi(x) := \left(\frac{\rho}{\varphi}\right)^p (x), \qquad x \in I,$$

we have (cf. (5.2.50))

$$\Phi \in Q_{\rho^p}(I) \quad \text{and} \quad \{x_k\}_{k=K_-}^{K_+} \in CS(\Phi, \rho^p, I, \alpha^p). \tag{5.2.62}$$

By (5.2.60) and (5.2.61),

$$B^p = \Big\| \int_{I_k} \Phi(t)\, d\bar{\nu} \Big\|_{\ell^{r/p}(\mathcal{K}+)},$$

and thus, by Remark 4.2.6 and (5.2.62),

$$B^p \approx \Big\| \frac{\Phi(x_k)}{\rho^p(x_k)} f(x_k) \Big\|_{\ell^{r/p}(\mathcal{K}_-^+)}, \tag{5.2.63}$$

where

$$f(x) := \|\min\{\rho^p(\cdot), \rho^p(x)\}\|_{1,I,\nu}, \quad x \in I, \tag{5.2.64}$$

(i.e., f is the ρ^p-fundamental function of the space $L^1(I,\nu)$). If

$$\{y_k\}_{k=K_-^1}^{K_+^1} \in CS(f, \rho^p, I, \beta) \quad \text{with} \quad \beta > 1,$$

then, by (5.2.63) and Lemma 4.2.9 (with $f_1 = \Phi$, $f_2 = f$), we obtain

$$B^p \approx \Big\| \frac{\Phi(y_k)}{\rho^p(y_k)} f(y_k) \Big\|_{\ell^{r/p}(\mathcal{K}_-^{1+})} \tag{5.2.65}$$

with

$$\mathcal{K}_-^{1+} := \{k \in \mathbb{Z};\, K_-^1 \le k \le K_+^1\}.$$

The equation (5.2.56) still holds, and thus, by Theorem 2.4.4, for all $x \in I$,

$$\begin{aligned} &\Big(\frac{\rho^q}{H}\Big)^{r/q}(x) \\ &\quad \approx \lim_{t\to a+} \Big(\frac{\rho^q}{H}\Big)^{r/q}(t) + \Big(\lim_{t\to b-} \frac{1}{H(t)}\Big)^{r/q} \rho^r(x) \\ &\quad + \int_I \frac{\min\{\rho^r(\cdot), \rho^r(x)\} \int_a^t \rho^q(s) w^q(s)\, ds \int_t^b w^q(s)\, ds}{H^{r/q+2}(t)}\, d\rho^q(t). \end{aligned} \tag{5.2.66}$$

This gives, for all $x \in I$,

$$\begin{aligned} \Phi(x) &= \Big(\frac{\rho}{\varphi}\Big)^p(x) = \Big[\Big(\frac{\rho^q}{H}\Big)^{r/q}(x)\Big]^{p/r} \\ &\approx \lim_{t\to a+} \Big(\frac{\rho}{\varphi}\Big)^p(t) + \Big(\lim_{t\to b-} \frac{1}{\varphi^p(t)}\Big) \rho^p(x) \\ &\quad + \|\min\{\rho^p(\cdot), \rho^p(x)\}\|_{r/p,W,I,\mu}, \end{aligned} \tag{5.2.67}$$

where

$$(5.2.68) \qquad W(t) := \frac{\left(\int_a^t \rho^q(s)\, w^q(s)\, ds\right)^{p/r} \left(\int_t^b w^q(s)\, ds\right)^{p/r}}{\varphi(t)^{p+2pq/r}}$$

and

$$(5.2.69) \qquad d\mu(t) := d\rho^q(t).$$

Thus, on using (5.2.67) in (5.2.65), we arrive at

$$(5.2.70) \qquad \begin{aligned} B^p \approx & \left(\lim_{t\to a+} \left(\frac{\rho}{\varphi}\right)^p(t)\right) \left\| \frac{f(y_k)}{\rho^p(y_k)} \right\|_{\ell^{r/p}(\mathcal{K}_-^{1+})} \\ & + \left(\lim_{t\to b-} \frac{1}{\varphi^p(t)}\right) \|f(y_k)\|_{\ell^{r/p}(\mathcal{K}_-^{1+})} \\ & + \left\| \|\min\{\rho^p(\cdot), \rho^p(y_k)\}\|_{r/p,W,I,\mu} \frac{f(y_k)}{\rho^p(y_k)} \right\|_{\ell^{r/p}(\mathcal{K}_-^{1+})} \\ =: & \, A_1 + A_2 + A_3 \end{aligned}$$

(recall that Convention 1.1.1 (i) is used when the expressions $0 \cdot (+\infty)$ appear in (5.2.70)). Consequently, condition (5.2.53) is equivalent to

$$(5.2.71) \qquad A_1^{1/p} + A_2^{1/p} + A_3^{1/p} + \frac{\psi^{1/p}(a)}{\varphi(a)} < +\infty.$$

Since the sequence $\{f(y_k)/\rho^p(y_k)\}_{k\in\mathcal{K}_-^{1+}}$ is geometrically decreasing and the sequence $\{f(y_k)\}_{k\in\mathcal{K}_-^{1+}}$ is geometrically increasing, it follows from Remark 1.3.2 and Lemma 1.3.3 that

$$A_1 \approx \lim_{t\to a+} \left(\frac{\rho}{\varphi}\right)^p(t) \lim_{t\to a+} \frac{f(t)}{\rho^p(t)} \qquad \text{and} \qquad A_2 \approx \lim_{t\to b-} \frac{1}{\varphi^p(t)} \lim_{t\to b-} f(t).$$

Moreover, using (5.2.7) and Lemma 2.3.2, we arrive at

$$\lim_{t\to a+} \left(\frac{\rho}{\varphi}\right)^p(t) = \frac{1}{\|w\|_{q,I}^p} \qquad \text{and} \qquad \lim_{t\to b-} \frac{1}{\varphi^p(t)} = \frac{1}{\|\rho w\|_{q,I}^p}.$$

Similarly, making use of (5.2.64), (5.2.55) and Lemma 2.3.2, we obtain

$$\lim_{t\to a+} \frac{f(t)}{\rho^p(t)} = \nu(I) = \int_I \rho^{-p}(t)\, d\|v\|_{p^*,(a,t)}^p$$

and

$$\lim_{t\to b-} f(t) = \|\rho^p\|_{1,I,\nu} = \int_I d\psi(t) = \int_I d\|v\|_{p^*,(a,t)}^p.$$

Consequently,

$$(5.2.72)\quad A_1^{1/p} \approx \frac{\left(\int_I \rho^{-p}(t)\, d\|v\|^p_{p^*,(a,t)}\right)^{1/p}}{\|w\|_{q,I}} \quad \text{and} \quad A_2^{1/p} \approx \frac{\left(\int_I d\|v\|^p_{p^*,(a,t)}\right)^{1/p}}{\|\rho w\|_{q,I}}.$$

If

$$\{z_k\}_{k=K^2_-}^{K^2_+} \in CS\left(\|\min\{\rho^p(\cdot), \rho^p(x)\}\|_{r/p,W,I,\mu}, \rho^p, I, \beta_1\right) \quad \text{with} \quad \beta_1 > 1,$$

then an application of Lemma 4.2.9 followed by Theorem 4.2.5 gives

$$(5.2.73)\quad A_3 \approx \left\| \|\min\{\rho^p(\cdot), \rho^p(z_k)\}\|_{r/p,W,I,\mu} \frac{f(z_k)}{\rho^p(z_k)} \right\|_{\ell^{r/p}(\mathcal{K}^{2+}_-)} \approx \|f\|_{r/p,W,I,\mu},$$

where

$$\mathcal{K}^{2+}_- := \{k \in \mathbb{Z}; K^2_- \le k \le K^2_+\}.$$

On collecting together (5.2.72) and (5.2.73), (5.2.68), (5.2.69) and (5.2.64), one can easily verify that condition (5.2.71) coincides with (5.2.4). □

Remark 5.2.7 If $q = +\infty$, then the fundamental function

$$(5.2.74)\qquad \varphi(x) = \|\min\{\rho(\cdot), \rho(x)\}\|_{\infty,w,I}, \quad x \in I,$$

is not in integral form and this is why the proof of Theorem 5.2.1 does not work in this case (one cannot apply Theorem 2.4.4 to get a representation of the function $(\rho/\varphi)^p$). However, as φ is a ρ-quasiconcave function, Theorem 2.4.1 asserts that there exists a non-negative Borel measure $\widetilde{\mu}$ such that

$$(5.2.75)\qquad \varphi(x) \approx \alpha + \beta\rho(x) + \int_I \min\{\rho(t), \rho(x)\}\, d\widetilde{\mu}(t) \quad \text{for all } x \in I.$$

Given α, β and $\widetilde{\mu}$, we obtain from Theorem 2.4.4, (5.2.75) and (5.2.74) that

$$(5.2.76)\qquad \left(\frac{\rho}{\varphi}\right)^p(x) \approx \alpha_1^p + \beta_1^p\rho^p(x) + \int_I \frac{\min\{\rho^p(t), \rho^p(x)\}}{\|\min\{\rho(\cdot), \rho(t)\}\|^{p+2}_{\infty,w,I}} \mathcal{V}(t)\, d\rho(t),$$

where

$$\alpha_1 = \lim_{t\to a+} \frac{\rho(t)}{\|\min\{\rho(\cdot), \rho(t)\}\|_{\infty,w,I}}, \qquad \beta_1 = \lim_{t\to b-} \frac{1}{\|\min\{\rho(\cdot), \rho(t)\}\|_{\infty,w,I}},$$

and, for all $t \in I$,

$$(5.2.77)\qquad \mathcal{V}(t) = \mathcal{V}(t; \alpha, \beta, \widetilde{\mu}, \rho) = \left(\alpha + \int_{(a,t]} \rho(s)\, d\widetilde{\mu}(s)\right)\left(\beta + \int_{[t,b)} d\widetilde{\mu}(s)\right).$$

Through use of the representation (5.2.76), we are able to establish the following theorem in the case $q = +\infty$; the proof follows along the same lines as that of Case B of Theorem 5.2.1.

Theorem 5.2.8 *Let $I = (a,b) \subseteq \mathbb{R}$, $\rho \in Ads(I)$ and $0 < p \le 1$. Let $v, w \in \mathcal{W}(I)$ and suppose that (5.2.1) is satisfied. Moreover, let the fundamental function $\varphi(x) = \|\min\{\rho(\cdot), \rho(x)\}\|_{\infty,w,I}$, $x \in I$, satisfy (5.2.75) for some non-negative Borel measure $\widetilde{\mu}$. Then*

$$\|g\|_{p,v,I} \lesssim \left\| \sup_{a<t\le x} \rho(t) \int_t^b g(s)\,ds \right\|_{\infty,w,I} \quad \text{for all } g \in \mathcal{M}^+(I), \tag{5.2.78}$$

if and only if

$$\left(\int_I \left(\int_I \min\left\{1, \frac{\rho^p(x)}{\rho^p(t)}\right\} d\|v\|^p_{p^*,(a,t)}\right) \frac{\mathcal{V}(x;\alpha,\beta,\widetilde{\mu},\rho)}{\|\min\{\rho(\cdot),\rho(x)\}\|^{p+2}_{\infty,w,I}}\, d\rho(x)\right)^{1/p} + \frac{\left(\int_I \rho^{-p}(t)\, d\|v\|^p_{p^*,(a,t)}\right)^{1/p}}{\|w\|_{\infty,I}} + \frac{\left(\int_I d\|v\|^p_{p^*,(a,t)}\right)^{1/p}}{\|\rho w\|_{\infty,w,I}} + \frac{\lim_{x\to a+} \|v\|_{p^*,(a,t)}}{\lim_{x\to a+} \|\min\{\rho(\cdot),\rho(x)\}\|_{\infty,w,I}} < +\infty, \tag{5.2.79}$$

where the function $\mathcal{V}(\,\cdot\,;\alpha,\beta,\widetilde{\mu},\rho)$ is given in (5.2.77).

The next lemma concerns the case when assumption (5.2.1) of Theorem 5.2.1 and Theorem 5.2.8 is violated. Note that in such a case the expression

$$\int_I \min\left\{1, \frac{\rho^p(x)}{\rho^p(t)}\right\} d\|v\|^p_{p^*,(a,t)}$$

appearing in (5.2.3), (5.2.4) and (5.2.79) does not make sense.

Lemma 5.2.9 *Let $I = (a,b) \subseteq \mathbb{R}$, $\rho \in Ads(I)$, $w, v \in W(I)$, $0 < p \le 1$ and $0 < q \le +\infty$. Assume that*

$$\|v\|_{p^*,(a,c)} = +\infty \text{ for some } c \in (a,b). \tag{5.2.80}$$

Then (5.2.2) holds if and only if

$$\varphi(x) := \|\min\{\rho(\cdot), \rho(x)\}\|_{q,w,I} = +\infty \quad \text{for all } x \in I. \tag{5.2.81}$$

Proof. Since

$$\|v\|_{p^*,(a,c)} = \sup_{g \in \mathcal{M}^+(a,c)} \frac{\|gv\|_{p,(a,c)}}{\|g\|_{1,(a,c)}},$$

we conclude from (5.2.80) that given $n \in \mathbb{N}$, there exists $\bar{g} = \bar{g}_n \in \mathcal{M}^+(a,c)$ such that

$$\|\bar{g}\|_{1,(a,c)} = 1 \quad \text{and} \quad \|\bar{g}v\|_{p,(a,c)} \ge n^{1/p}. \tag{5.2.82}$$

Define the function $g = g_n \in \mathcal{M}^+(I)$ to be the extension of $\bar{g}$ by zero in $[c, b)$. Then, by (5.2.82),

$$\text{LHS (5.2.2)} = \|gv\|_{p,(a,b)} = \|\bar{g}v\|_{p,(a,c)} \geq n^{1/p}, \tag{5.2.83}$$

$$\|g\|_{1,(a,c)} = \|g\|_{1,I} = 1. \tag{5.2.84}$$

Assuming that (5.2.2) holds on $\mathcal{M}^+(I)$, we obtain from (5.2.83) that RHS (5.2.2) $\gtrsim n^{1/p}$. Thus, using also (5.2.84) and the fact that $\rho \in \mathcal{M}^+(I; \uparrow)$, we arrive at

$$\begin{aligned}
n^{1/p} &\lesssim \Big\| \sup_{a<t\leq x} \rho(t) \int_t^b g(s)\,ds \Big\|_{q,w,I} \\
&\approx \Big\| \sup_{a<t\leq x} \rho(t) \int_t^b g(s)\,ds \Big\|_{q,w,(a,c)} + \Big\| \sup_{a<t\leq x} \rho(t) \int_t^b g(s)\,ds \Big\|_{q,w,(c,b)} \\
&\leq \|\rho(\cdot)\|_{q,w,(a,c)} + \Big\| \sup_{a<t\leq c} \rho(t) \int_t^b g(s)\,ds \Big\|_{q,w,(c,b)} \\
&\qquad + \Big\| \sup_{c<t\leq x} \rho(t) \int_t^b g(s)\,ds \Big\|_{q,w,(c,b)} \\
&\leq \|\rho(\cdot)\|_{q,w,(a,c)} + \|\rho(c)\|_{q,w,(c,b)} \\
&\approx \|\min\{\rho(\cdot), \rho(c)\}\|_{q,w,I} \\
&= \varphi(c).
\end{aligned}$$

Since $n \in \mathbb{N}$ was arbitrary, we deduce that $\varphi(c) = +\infty$. However, this implies that $\varphi \equiv +\infty$ on I.

On the other hand, if $\varphi \equiv +\infty$ on I, then it was shown in the proof of Theorem 5.2.1 that (5.2.2) holds. □

5.3 The inequality $\Big\| \int_a^x u(t) \Big(\int_t^b g(s)\,ds \Big)\,dt \Big\|_{q,w,I} \lesssim \|g\|_{p,v,I}$

The next theorem is an analogue of Theorem 5.1.1 involving another ρ-quasiconcave operator.

Theorem 5.3.1 *Let $I = (a, b) \subseteq \mathbb{R}$, $1 \leq p \leq +\infty$ and $0 < q \leq +\infty$. Let $w, v, u \in \mathcal{W}(I)$ and suppose that $\rho(x) := \int_a^x u(t)\,dt$, $x \in I$, is such that $\rho \in Ads(I)$. Then*

$$\Big\| \int_a^x u(t) \Big(\int_t^b g(s)\,ds \Big)\,dt \Big\|_{q,w,I} \lesssim \|g\|_{p,v,I} \quad \textit{for all } g \in \mathcal{M}^+(I), \tag{5.3.1}$$

if and only if one of the following is satisfied:

Case A: $p \le q$ *and*

$$\left\| \left\| \min\left\{ \frac{\rho(\cdot)}{\rho(x)}, 1 \right\} \right\|_{q,w,I} \left\| \min\{\rho(\cdot), \rho(x)\} \right\|_{p',v^{-1},I} \right\|_{\infty,I} < +\infty; \tag{5.3.2}$$

Case B: $p > q$ *and*

$$\left\| \left\| \min\left\{ \frac{\rho(\cdot)}{\rho(x)}, 1 \right\} \right\|_{q,w,I}^{1-q/r} w^{q/r}(x) \left\| \min\{\rho(\cdot), \rho(x)\} \right\|_{p',v^{-1},I} \right\|_{r,I} < +\infty, \tag{5.3.3}$$

where $1/r := 1/q - 1/p$.

We follow the same pattern as for the proofs of Theorems 5.1.1 and 5.2.1 by beginning with results which apply to both cases of the theorem in order to clarify the exposition. The first is the analogue of Lemma 5.1.3.

Lemma 5.3.2 *Let* $I = (a,b) \subseteq \mathbb{R}$, $1 \le p \le +\infty$ *and* $0 < q \le +\infty$. *Suppose that* $w,\ v,\ u \in \mathcal{W}(I)$, $\rho(x) := \int_a^x u(t)dt\ \in Ads(I)$, *and that*

$$\varphi(x) := \|\min\{\rho(\cdot), \rho(x)\}\|_{q,w,I} < +\infty \quad \text{for all } x \in I. \tag{5.3.4}$$

Let

$$\{x_k\}_{k=K_-}^{K_+} \in CS(\varphi, \rho, I, \alpha) \quad \text{with } \alpha > 2^{1/q}, \tag{5.3.5}$$

and define [5]

$$A_k := \sup_{h \in \mathcal{M}^+(I_k)} \left(\int_{I_k} \varphi(t) h(t)\, dt \right) \Big/ \|hv\|_{p,I_k}, \quad k \in \mathcal{K}^+, \tag{5.3.6}$$

where

$$I_k = (x_{k-1}, x_k]. \tag{5.3.7}$$

Then (5.3.1) *holds if and only if*

$$\|\{A_k\}\|_{\ell^r(\mathcal{K}^+)} < +\infty, \tag{5.3.8}$$

where

$$\frac{1}{r} := \left(\frac{1}{q} - \frac{1}{p} \right)_+. \tag{5.3.9}$$

Proof. By Remark 2.2.5, the function $f \in Q_\rho(I)$ given by

$$f(x) := \int_a^x u(t) \left(\int_t^b g(s) ds \right) dt, \quad \text{where} \quad g \in \mathcal{M}^+(I), \tag{5.3.10}$$

[5] Note that in (5.3.6) we use the convention that $0/0 = 0$ and $(+\infty)/(+\infty) = 0$ - cf. Convention 1.1.1 (i).

can be rewritten as

$$f(x) = \| \min\{\rho(\cdot), \rho(x)\}\|_{1,g,I}, \; x \in I.$$

Thus, by Theorem 4.2.5,

$$\text{LHS}(5.3.1) \approx \left\| \int_{I_k} \varphi(t) g(t)\, dt \right\|_{\ell^q(\mathcal{K}^+)}. \tag{5.3.11}$$

Since

$$\text{RHS }(5.3.1) = \| \, \|gv\|_{p,I_k} \|_{\ell^p(\mathcal{K}^+)}, \tag{5.3.12}$$

inequality (5.3.1) can be rewritten as

$$\left\| \int_{I_k} \varphi(t) g(t)\, dt \right\|_{\ell^q(\mathcal{K}^+)} \lesssim \| \, \|gv\|_{p,I_k} \|_{\ell^p(\mathcal{K}^+)} \quad \text{for all } g \in \mathcal{M}^+(I). \tag{5.3.13}$$

The associated "local" Hardy-type inequality, for each $k \in \mathcal{K}^+$, is

$$\int_{I_k} \varphi(t) g(t)\, dt \le A_k \, \|gv\|_{p,I_k}, \quad g \in \mathcal{M}^+(I),$$

with best possible constant A_k given by (5.3.6). The rest of the proof is similar to that of Lemma 5.1.3. □

Lemma 5.3.3 *Suppose that the conditions of* Lemma 5.3.2 *are satisfied. Then* (5.3.8), *and hence* (5.3.1), *is equivalent to*

$$\|\{\|\varphi v^{-1}\|_{p',I_k}\}\|_{\ell^r(\mathcal{K}^+)} < +\infty. \tag{5.3.14}$$

Proof. This follows from the fact that the application of Hölder's inequality in

$$\int_{I_k} \varphi(t) h(t)\, dt \le \|\varphi(t) v^{-1}(t)\|_{p',I_k} \|hv\|_{p,I_k}$$

is sharp, and hence that

$$A_k = \|\varphi(t) v^{-1}(t)\|_{p',I_k}. \tag{5.3.15}$$

The lemma is therefore an immediate consequence of Lemma 5.3.2.

□

Proof of Theorem 5.3.1. The proof involves applying the antidiscretization method to (5.3.14). If

$$\varphi(x) = +\infty \quad \text{for some} \quad x \in I,$$

the theorem is proved by a similar argument to that in the proof of Theorem 5.1.1, but now with

$$f(x) = \int_a^x u(t) \left(\int_t^b g(s)ds \right) dt, \quad \text{where } g \in \mathcal{M}^+(I).$$

Therefore, we may suppose that

$$\varphi(x) < +\infty \quad \text{for all} \quad x \in I. \tag{5.3.16}$$

Case A: In this case, the discrete condition (5.3.14) becomes

$$\|\{\|\varphi v^{-1}\|_{p',I_k}\}\|_{\ell^\infty(\mathcal{K}^+)} < +\infty. \tag{5.3.17}$$

It is necessary to consider separate cases within Case A.
Case A-1: $1 < p \le q \le +\infty$. Then $p' < +\infty$ and (5.3.17) is equivalent to

$$\left\| \int_{I_k} \varphi^{p'}(t)\, v^{-p'}(t)\, dt \right\|_{\ell^\infty(\mathcal{K}^+)} < +\infty. \tag{5.3.18}$$

Even in this case we shall distinguish two cases.
Case A-1-1: $1 < p \le q = +\infty$. Then (cf. (5.3.4))

$$\varphi^{p'}(x) = \|\min\{\rho^{p'}(\cdot), \rho^{p'}(x)\} w^{p'}(\cdot)\|_{\infty,I}, \quad x \in I,$$

i.e.,

$$\Phi(x) := \varphi^{p'}(x) = \|\min\{R(\cdot), R(x)\}\|_{\infty,W,I}, \quad x \in I,$$

where

$$R = \rho^{p'} \quad \text{and} \quad W = w^{p'}. \tag{5.3.19}$$

Moreover, by (5.3.16),

$$\Phi(x) < +\infty \qquad \text{for all } x \in I.$$

As also (cf. Remark 3.2.3 (ii))

$$\{x_k\}_{k=K_-}^{K_+} \in CS(\Phi, R, I, \alpha^{p'}), \quad \alpha > 2^{1/q},$$

Theorem 4.2.5 (with $q = +\infty$) implies that (5.3.18) is equivalent to

$$\|\|\min\{R(\cdot), R(x)\}\|_{1,v^{-p'},I}\|_{\infty,W,I} < +\infty, \tag{5.3.20}$$

which can be rewritten as

$$\|(\|\min\{R(\cdot), R(x)\}\|_{1,v^{-p'},I})^{1/p'} W^{1/p'}(x)\|_{\infty,I} < +\infty. \tag{5.3.21}$$

On using (5.3.19), one can see that (5.3.21) is equivalent to

$$\|\,\|\min\{\rho(\cdot), \rho(x)\}\|_{p',v^{-1},I}\|_{\infty,w,I} < +\infty. \tag{5.3.22}$$

Since the function

$$h(x) := \| \min\{\rho(\cdot), \rho(x)\}\|_{p',v^{-1},I}, \quad x \in I,$$

belongs to the class $Q_\rho(I)$, the desired result (5.3.2) follows from (5.3.22) and Lemma 2.2.14.

Case A-1-2: $1 < p \le q < +\infty$. Then, by (5.3.4),

$$\varphi^q(x) = \int_I \min\{\rho^q(t), \rho^q(x)\} w^q(t)\, dt, \quad x \in I, \tag{5.3.23}$$

and, on applying Theorem 2.4.3,

$$\begin{aligned} \varphi^{p'}(x) &= (\varphi^q(x))^{p'/q} \\ &\approx \left\{ \int_I \min\left\{ (\rho^q(t))^{p'/q}, (\rho^q(x))^{p'/q} \right\} \left(\frac{\varphi^q}{\rho^q}\right)^{p'/q-1}(t)\, w^q(t)\, dt \right. \\ &= \| \min\{\rho^{p'}(\cdot), \rho^{p'}(x)\}\|_{1,\omega,I}, \quad x \in I, \end{aligned} \tag{5.3.24}$$

where

$$\omega = \left(\frac{\varphi^q}{\rho^q}\right)^{p'/q-1} w^q. \tag{5.3.25}$$

Consequently, by Lemma 2.4.5,

$$\varphi^{p'}(x) \approx \| \min\{\rho^{p'}(\cdot), \rho^{p'}(x)\}\|_{\infty,W_\infty,I} \quad \text{for all } x \in I,$$

where

$$W_\infty = \frac{\varphi^{p'}}{\rho^{p'}}. \tag{5.3.26}$$

Hence,

$$\Phi(x) := \varphi^{p'}(x) \approx \| \min\{R(\cdot), R(x)\}\|_{\infty,W,I} =: \widetilde{\Phi}(x) \quad \text{for all } x \in I, \tag{5.3.27}$$

with

$$R = \rho^{p'} \quad \text{and} \quad W = W_\infty. \tag{5.3.28}$$

Condition (5.3.18) has the form

$$\left\| \int_{I_k} \Phi(t)\, v^{-p'}(t)\, dt \right\|_{\ell^\infty(\mathcal{K}^+)} < +\infty. \tag{5.3.29}$$

Moreover, by (5.3.16) and (5.3.27),

$$\Phi(x) < +\infty \quad \text{and} \quad \widetilde{\Phi}(x) < +\infty \qquad \text{for all } x \in I. \tag{5.3.30}$$

Firstly, by Remark 4.2.6, condition (5.3.29) is equivalent to

$$\left\| \frac{\Phi(x_k)}{R(x_k)} \| \min\{R(\cdot), R(x_k)\}\|_{1,v^{-p'},I} \right\|_{\ell^\infty(\mathcal{K}_-^+)} < +\infty. \tag{5.3.31}$$

Secondly, Corollary 4.2.10 (with functions $f_1 := \Phi$, $f_2 := \widetilde{\Phi}$, $\rho := R$ and $F(x) := \| \min\{R(\cdot), R(x)\}\|_{1,v^{-p'},I}$, $x \in I$) implies that condition (5.3.31) is equivalent to

$$\left\| \frac{\widetilde{\Phi}(\widetilde{x}_k)}{R(\widetilde{x}_k)} \| \min\{R(\cdot), R(\widetilde{x}_k)\}\|_{1,v^{-p'},I} \right\|_{\ell^\infty(\widetilde{\mathcal{K}}_-^+)} < +\infty, \tag{5.3.32}$$

where $\{\widetilde{x}_k\}_{k=\widetilde{K}_-}^{\widetilde{K}_+}$ is any covering sequence satisfying

$$\{\widetilde{x}_k\}_{k=\widetilde{K}_-}^{\widetilde{K}_+} \in CS(\widetilde{\Phi}, R, I, \beta) \quad \text{with } \beta > 1, \tag{5.3.33}$$

and

$$\widetilde{\mathcal{K}}_-^+ := \{k \in \mathbb{Z}; \widetilde{K}_- \le k \le \widetilde{K}_+\}. \tag{5.3.34}$$

Finally, we apply Theorem 4.2.5 to show that condition (5.3.32) is equivalent to (5.3.20), which can be rewritten as (5.3.21). On using (5.3.28), (5.3.26) and (5.3.4), one can see that (5.3.21) coincides with (5.3.2).

Case A-2: $1 = p \le q \le +\infty$. In this case we have $p' = \infty$, and (5.3.17) becomes

$$\| \, \|\varphi v^{-1}\|_{\infty,I_k} \|_{\ell^\infty(\mathcal{K}^+)} < +\infty. \tag{5.3.35}$$

Also in this case it is convenient to distinguish two cases.

Case A-2-1: $1 = p$, $q = +\infty$. Then

$$\varphi(x) = \| \min\{\rho(\cdot), \rho(x)\}\|_{\infty,w,I}, \quad x \in I,$$

and Remark 4.2.4 (i) shows that condition (5.3.35) is equivalent to (5.3.22) (since now $p' = +\infty$), which, together with Lemma 2.2.14, gives (5.3.2).

Case A-2-2: $1 = p \le q < +\infty$. Then, by (5.3.4), we have (5.3.23), and Theorem 2.4.3 implies that

$$\begin{aligned} \varphi(x) = (\varphi^q(x))^{1/q} &\approx \int_I \min\{\rho(t), \rho(x)\} \left(\frac{\varphi^q}{\rho^q} \right)^{1/q-1} (t)\, w^q(t)\, dt \\ &= \| \min\{\rho(\cdot), \rho(x)\|_{1,\omega,I}, \quad x \in I, \end{aligned} \tag{5.3.36}$$

where

$$\omega = \left(\frac{\varphi^q}{\rho^q} \right)^{1/q-1} w^q. \tag{5.3.37}$$

Consequently, by Lemma 2.4.5,

$$\varphi(x) \approx \| \min\{\rho(\cdot), \rho(x)\}\|_{\infty,W,I} =: \widetilde{\Phi}(x) \quad \text{for all } x \in I, \tag{5.3.38}$$

with

$$W = \frac{\varphi}{\rho}. \tag{5.3.39}$$

Moreover, by (5.3.16) and (5.3.38),

$$\varphi(x) < +\infty \quad \text{and} \quad \widetilde{\Phi}(x) < +\infty \qquad \text{for all } x \in I. \tag{5.3.40}$$

Firstly, by Remark 4.2.4 (ii), Theorem 4.2.3 and Remark 4.2.4 (i), condition (5.3.35) is equivalent to

$$\left\| \frac{\varphi(x_k)}{\rho(x_k)} \| \min\{\rho(\cdot), \rho(x_k)\}\|_{\infty,v^{-1},I} \right\|_{\ell^\infty(\mathcal{K}^+_-)}. \tag{5.3.41}$$

Secondly, Corollary 4.2.10 and (5.3.38) imply that (5.3.41) is equivalent to

$$\left\| \frac{\widetilde{\Phi}(\widetilde{x}_k)}{\rho(\widetilde{x}_k)} \| \min\{\rho(\cdot), \rho(\widetilde{x}_k)\}\|_{\infty,v^{-1},I} \right\|_{\ell^\infty(\widetilde{\mathcal{K}}^+_-)}, \tag{5.3.42}$$

where $\{\widetilde{x}_k\}_{k=\widetilde{K}_-}^{\widetilde{K}^+}$ is any covering sequence satisfying (5.3.33), and $\widetilde{\mathcal{K}}^+_-$ is given by (5.3.34). Finally, Remark 4.2.4 (i) shows that (5.3.42) is equivalent to

$$\| \, \| \min\{\rho(\cdot), \rho(x)\}\|_{\infty,v^{-1},I} \, \|_{\infty,W,I} < +\infty,$$

which, together with (5.3.39), gives the desired result (5.3.2).
Case B: In this case the discrete condition (5.3.14) becomes

$$\|\{\|\varphi v^{-1}\|_{p',I_k}\}\|_{\ell^r(\mathcal{K}^+)} < +\infty \tag{5.3.43}$$

with $1/r = 1/q - 1/p$.

It is necessary to consider separate cases.
Case B-1: $1 < p \le +\infty$, $0 < q < p$. Then $p' < +\infty$ and (5.3.43) can be rewritten as

$$\left\| \int_{I_k} \varphi^{p'}(t)\, v^{-p'}(t)\, dt \right\|_{\ell^{r/p'}(\mathcal{K}^+)} < +\infty. \tag{5.3.44}$$

From (5.3.24), (5.3.25) and Lemma 2.4.5, we obtain

$$\Phi(x) := \varphi^{p'}(x) \approx \| \min\{R(\cdot), R(x)\}\|_{r/p',W,I} =: \widetilde{\Phi}(x) \quad \text{for all } x \in I, \tag{5.3.45}$$

where

$$R = \rho^{p'} \quad \text{and} \quad W = \left(\frac{\varphi}{\rho}\right)^{p'} \left(\left(\frac{\rho}{\varphi}\right)^{p'} \omega \right)^{p'/r} = \left(\frac{\varphi}{\rho}\right)^{p'} \left(\frac{\rho}{\varphi} w\right)^{qp'/r}. \tag{5.3.46}$$

Moreover, by (5.3.16) and (5.3.45), (5.3.30) holds. Firstly, by Remark 4.2.6, condition (5.3.44) is equivalent to

$$\left\| \frac{\Phi(x_k)}{R(x_k)} \| \min\{R(\cdot), R(x_k)\}\|_{1,v^{-p'},I} \right\|_{\ell^{r/p'}(\mathcal{K}_-^+)} < +\infty. \tag{5.3.47}$$

Secondly, condition (5.3.45) and Corollary 4.2.10 imply that condition (5.3.47) is equivalent to

$$\left\| \frac{\widetilde{\Phi}(\widetilde{x}_k)}{R(\widetilde{x}_k)} \| \min\{R(\cdot), R(\widetilde{x}_k)\}\|_{1,v^{-p'},I} \right\|_{\ell^{r/p'}(\widetilde{\mathcal{K}}_-^+)} < +\infty, \tag{5.3.48}$$

where $\{\widetilde{x}_k\}_{k=\widetilde{K}_-}^{\widetilde{K}_+}$ is any covering sequence satisfying (5.3.33) and $\widetilde{\mathcal{K}}_-^+$ is given by (5.3.34). Finally, we apply Theorem 4.2.5 to show that condition (5.3.48) is equivalent to

$$\| \| \min\{R(\cdot), R(x)\}\|_{1,v^{-p'},I}\|_{r/p',W,I} < +\infty, \tag{5.3.49}$$

which can be rewritten as

$$\|(\| \min\{R(\cdot), R(x)\}\|_{1,v^{-p'},I})^{1/p'} W^{1/p'}\|_{r,I} < +\infty. \tag{5.3.50}$$

On using (5.3.46) and (5.3.4), one can see that (5.3.50) coincides with (5.3.3).
Case B-2: $p = 1,\ 0 < q < p$. Then $p' = +\infty$ and (5.3.43) becomes

$$\|\{\|\varphi v^{-1}\|_{\infty,I_k}\}\|_{\ell^r(\mathcal{K}^+)} < +\infty. \tag{5.3.51}$$

From (5.3.23), and Theorem 2.4.3, we obtain

$$\begin{aligned} \varphi(x) = (\varphi^q(x))^{1/q} &\approx \int_I \min\{\rho(t), \rho(x)\} \left(\tfrac{\varphi^q}{\rho^q}\right)^{1/q-1}(t)\, w^q(t)\, dt \\ &= \| \min\{\rho(\cdot), \rho(x)\}\|_{1,\omega,I} \quad \text{for all } x \in I, \end{aligned}$$

where

$$\omega = \left(\frac{\varphi^q}{\rho^q}\right)^{1/q-1} w^q.$$

Lemma 2.4.5 then gives

$$\varphi(x) \approx \| \min\{\rho(\cdot), \rho(x)\}\|_{r,W,I} =: \widetilde{\Phi}(x) \quad \text{for all } x \in I, \tag{5.3.52}$$

where

$$W = \left(\frac{\varphi}{\rho}\right)\left(\frac{\rho\,\omega}{\varphi}\right)^{1/r} = \left(\frac{\varphi}{\rho}\right)^{1-q/r} w^{q/r}. \tag{5.3.53}$$

Moreover, by (5.3.16) and (5.3.52), (5.3.40) holds. Firstly, by Remark 4.2.4 (iii), Theorem 4.2.3 and Remark 4.2.4 (i), condition (5.3.51) is equivalent to

$$\left\| \frac{\varphi(x_k)}{\rho(x_k)} \| \min\{\rho(\cdot), \rho(x_k)\}\|_{\infty,v^{-1},I} \right\|_{\ell^r(\mathcal{K}_-^+)} < +\infty. \tag{5.3.54}$$

Secondly, condition (5.3.52) and Corollary 4.2.10 imply that condition (5.3.54) is equivalent to

$$(5.3.55)\qquad \left\| \frac{\widetilde{\Phi}(\widetilde{x}_k)}{\rho(\widetilde{x}_k)} \| \min\{\rho(\cdot), \rho(\widetilde{x}_k)\}\|_{\infty, v^{-1}, I} \right\|_{\ell^r(\widetilde{\mathcal{K}}_-^+)} < +\infty,$$

where $\{\widetilde{x}_k\}_{k=\widetilde{K}_-}^{\widetilde{K}_+}$ is any covering sequence satisfying (5.3.33) with $R = \rho$ and $\widetilde{\mathcal{K}}_-^+$ is given by (5.3.34). Finally, Remark 4.2.4 (i) shows that (5.3.55) is equivalent to

$$\| \| \min\{\rho(\cdot), \rho(x)\}\|_{\infty, v^{-1}, I} \|_{r, W, I} < +\infty,$$

which, together with (5.3.53), gives the desired result (5.3.3). □

5.4 The inequality $\|g\|_{p,v,I} \lesssim \left\| \int_a^x u(t) \left(\int_t^b g(s)\, ds \right) dt \right\|_{q,w,I}$

The following theorem concerns the reverse of the Hardy-type inequality (5.3.1) and is an analogue of Theorem 5.2.1.

Theorem 5.4.1 *Let $I = (a,b) \subseteq \mathbb{R}$, $w, v, u \in \mathcal{W}(I), 0 < p \le 1$ and $0 < q < +\infty$. Suppose that $\rho(x) := \int_a^x u(t)\, dt$, $x \in I$, is such that $\rho \in Ads(I)$. Then*

$$(5.4.1)\qquad \|g\|_{p,v,I} \lesssim \left\| \int_a^x u(t) \left(\int_t^b g(s)\, ds \right) dt \right\|_{q,w,I} \quad \text{for all } g \in \mathcal{M}^+(I),$$

if and only if one of the following is satisfied:
Case A: $q \le p$ *and*

$$(5.4.2)\qquad \left\| \left\| \min\left\{1, \frac{\rho(x)}{\rho(\cdot)}\right\} \right\|_{p^*, v, I} \| \min\{\rho(\cdot), \rho(x)\}\|_{q,w,I}^{-1} \right\|_{\infty, I} < +\infty;$$

Case B: $p < q$ *and*

$$(5.4.3)\qquad \left(\int_I \left\| \min\left\{1, \frac{\rho(x)}{\rho(\cdot)}\right\} \right\|_{p^*, v, I}^r \frac{\int_a^x \rho^q(s) w^q(s)\, ds \int_x^b w^q(s)\, ds}{\| \min\{\rho(\cdot), \rho(x)\}\|_{q,w,I}^{r+2q}} \, d\rho^q(x) \right)^{1/r} + \frac{\|v/\rho\|_{p^*, I}}{\|w\|_{q,I}} + \frac{\|v\|_{p^*, I}}{\|\rho w\|_{q,I}} < +\infty,$$

where $1/r := 1/p - 1/q$.

To prove Theorem 5.4.1, we need two lemmas.

Lemma 5.4.2 *Let $I = (a,b) \subseteq \mathbb{R}$, $w, v, u \in \mathcal{W}(I), 0 < p \le 1$ and $0 < q \le +\infty$. Suppose that $\rho(x) := \int_a^x u(t)\, dt$, $x \in I$, is such that $\rho \in Ads(I)$ and that*

$$(5.4.4) \qquad \varphi(x) := \| \min\{\rho(\cdot), \rho(x)\}\|_{q,w,I} < +\infty \quad \textit{for all} \quad x \in I.$$

Let

$$(5.4.5) \qquad \{x_k\}_{k=K_-}^{K_+} \in CS(\varphi, \rho, I, \alpha) \quad \textit{with} \quad \alpha > 2^{1/q},$$

and define

$$(5.4.6) \qquad B_k := \sup_{h \in \mathcal{M}^+(I)} \|hv\|_{p,I_k} \Big/ \int_{I_k} \varphi(t) h(t)\, dt, \quad k \in \mathcal{K}^+,$$

where $I_k = (x_{k-1}, x_k]$. [6] *Then (5.4.1) is equivalent to*

$$(5.4.7) \qquad \|\{B_k\}\|_{\ell^r(\mathcal{K}^+)} < +\infty,$$

where

$$(5.4.8) \qquad \frac{1}{r} = \left(\frac{1}{p} - \frac{1}{q} \right)_+ .$$

Proof. By Remark 2.2.5, the function

$$(5.4.9) \qquad f(x) := \int_a^x u(t) \left(\int_t^b g(s)\, ds \right) dt, \quad \text{where} \quad g \in \mathcal{M}^+(I),$$

can be rewritten as

$$f(x) = \| \min\{\rho(\cdot), \rho(x)\}\|_{1,g,I}, \quad x \in I.$$

Thus, by Theorem 4.2.5,

$$(5.4.10) \qquad \text{RHS (5.4.1)} \approx \left\| \int_{I_k} \varphi(t) g(t)\, dt \right\|_{\ell^q(\mathcal{K}^+)},$$

where

$$(5.4.11) \qquad I_k = (x_{k-1}, x_k].$$

Since

$$(5.4.12) \qquad \text{LHS (5.4.1)} = \| \, \|gv\|_{p,I_k} \|_{\ell^p(\mathcal{K}^+)},$$

inequality (5.4.1) can be rewritten as

$$(5.4.13) \qquad \| \, \|gv\|_{p,I_k} \|_{\ell^p(\mathcal{K}^+)} \lesssim \left\| \int_{I_k} \varphi(t) g(t)\, dt \right\|_{\ell^q(\mathcal{K}^+)} \quad \text{for all} \quad g \in \mathcal{M}^+(I).$$

---

[6] Note that in (5.4.6) we use the convention that $0/0 = 0$ and $(+\infty)/(+\infty) = 0$ - cf. Convention 1.1.1 (i).

The associated "local" reverse Hardy-type inequality, for each $k \in \mathcal{K}^+$, is

$$\|gv\|_{p,I_k} \le B_k \int_{I_k} \varphi(t) g(t)\, dt, \quad g \in \mathcal{M}^+, \tag{5.4.14}$$

with the best possible constant B_k given by (5.4.6). The rest of the proof now follows the familiar lines of the proofs of Lemmas 5.2.3 and 5.1.3.

□

Lemma 5.4.3 *Suppose that the conditions of* Lemma 5.4.2 *are satisfied. Then* (5.4.7), *and hence* (5.4.1), *is equivalent to*

$$B := \Big\| \, \|\varphi^{-1} v\|_{p^*,I_k} \Big\|_{\ell^r(\mathcal{K}^+)} < +\infty. \tag{5.4.15}$$

Proof. The sharpness of Hölder's inequality

$$\|gv\|_{p,I_k}^p \le \|\varphi g\|_{1,I_k}^p \, \|\varphi^{-1} v\|_{p^*,I_k}^p,$$

and the definition of B_k in (5.4.6), imply that

$$B_k = \|\varphi^{-1} v\|_{p^*,I_k}. \tag{5.4.16}$$

The lemma follows on applying this fact in Lemma 5.4.2.

□

Proof of Theorem 5.4.1. The case

$$\varphi(x) = +\infty \quad \text{for some} \quad x \in I$$

follows as in previous theorems, now with the function f given in (5.4.9). Therefore, we may assume that

$$\varphi(x) < +\infty \quad \text{for all} \quad x \in I. \tag{5.4.17}$$

We are required to verify that, under this assumption, the "discrete condition" (5.4.15) with $r = \infty$ is equivalent to the "continuous criterion" (5.4.2) in Case A, and that (5.4.15) with $1/r = 1/p - 1/q$ is equivalent to the "continuous criterion" (5.4.3) in Case B. To antidiscretise (5.4.15), we again distinguish several cases.

Case A: $q \le p$. Suppose first that $p < 1$ and thus $p^* < +\infty$. Then condition (5.4.15) can be written as

$$B^{p*} = \Big\| \int_{I_k} \Big(\frac{\rho}{\varphi}\Big)^{p^*}(t)\, d\bar{\nu} \Big\|_{\ell^\infty(\mathcal{K}^+)} < +\infty, \tag{5.4.18}$$

where ν is the Borel measure given by

$$d\nu := \rho^{-p^*}(t)\, v^{p^*}(t)\, dt, \tag{5.4.19}$$

and $\bar{\nu}$ is its extension by zero in $\mathbb{R} \setminus I$. If we find $R \in Ads(I)$ and $W \in \mathcal{W}(I)$ such that

$$(5.4.20) \qquad \left(\frac{\rho}{\varphi}\right)^{p^*}(x) \approx \| \min\{R(\cdot), R(x)\}\|_{\infty,W,I} =: \widetilde{\Phi}(x) \quad \text{for all } x \in I,$$

then, by Theorem 4.2.5 (with $q = +\infty$), Remark 4.2.6, and the usual application of Corollary 4.2.10, condition (5.4.18) will be equivalent to

$$(5.4.21) \qquad B^{p*} \approx \| \, \| \min\{R(\cdot), R(x)\}\|_{1,I,\nu}\|_{\infty,W,I} < +\infty.$$

We have from (5.4.4) that

$$(5.4.22) \qquad \varphi^q(x) = \int_I \min\{\rho^q(t), \rho^q(x)\}\, w^q(t)\, dt =: H(x), \quad x \in I.$$

Thus, by Lemma 2.4.6,

$$(5.4.23) \qquad \left(\frac{\rho}{\varphi}\right)^q(x) \approx \| \min\{\rho^q(\cdot), \rho^q(x)\}\|_{\infty,1/H,I} \quad \text{for all } x \in I,$$

which yields

$$(5.4.24) \qquad \left(\frac{\rho}{\varphi}\right)^{p^*}(x) = \left[\left(\frac{\rho}{\varphi}\right)^q(x)\right]^{p^*/q}$$
$$\approx \| \min\{\rho^{p^*}(\cdot), \rho^{p^*}(x)\}\|_{\infty,H^{-p^*/q},I} \quad \text{for all } x \in I.$$

Hence, (5.4.20) holds with

$$(5.4.25) \qquad R = \rho^{p^*} \quad \text{and} \quad W = H^{-p^*/q},$$

and (5.4.21) is verified.

Using (5.4.25), we can rewrite (5.4.21) as

$$B \approx \left\| \left[\int_I \min\{\rho^{p^*}(t), \rho^{p^*}(x)\}\, d\nu \right]^{1/p^*} H^{-1/q}(x) \right\|_{\infty,I} < +\infty,$$

which, with respect to (5.4.19) and (5.4.22), coincides with (5.4.2).

Let now $p = 1$ and thus $p^* = \infty$. Then condition (5.4.15) takes the form

$$(5.4.26) \qquad B := \| \, \|\varphi^{-1} v\|_{\infty,I_k}\|_{\ell^\infty(\mathcal{K}^+)} = \left\| \, \left\| \frac{\rho}{\varphi} \right\|_{\infty,v/\rho,I_k} \right\|_{\ell^\infty(\mathcal{K}^+)} < +\infty.$$

We follow a familiar path. If we are able to find $R \in Ads(I)$ and $W \in \mathcal{W}(I)$ such that

$$(5.4.27) \qquad \left(\frac{\rho}{\varphi}\right)(x) \approx \| \min\{R(\cdot), R(x)\}\|_{\infty,W,I} \quad \text{for all } x \in I,$$

then, by Remarks 4.2.4, Theorem 4.2.3 and Corollary 4.2.10, condition (5.4.26) would be equivalent to

$$(5.4.28) \qquad B \approx \| \, \| \min\{R(\cdot), R(x)\}\|_{\infty,v/\rho,I}\|_{\infty,W,I} < +\infty.$$

By (5.4.23),

$$\left(\frac{\rho}{\varphi}\right)(x) = \left[\left(\frac{\rho}{\varphi}\right)^q\right]^{1/q}(x) \approx \|\min\{\rho(\cdot), \rho(x)\}\|_{\infty, H^{-1/q}, I} \quad \text{for all } x \in I.$$

Consequently, (5.4.27) holds with

$$R = \rho \quad \text{and} \quad W = H^{-1/q} = \varphi^{-1}. \tag{5.4.29}$$

It follows that (5.4.28) is equivalent to (5.4.2) in this case also, and Case A of the theorem is proved.

Case B: $p < q$. Suppose first that $p < 1$ and so $p^* < +\infty$. Then condition (5.4.15) is equivalent to

$$B^{p^*} = \left\| \int_{I_k} \left(\frac{\rho}{\varphi}\right)^{p^*}(t)\, d\overline{\nu} \right\|_{\ell^{r/p^*}(\mathcal{K}+)} < +\infty, \tag{5.4.30}$$

where $1/r = 1/p - 1/q$ and $\overline{\nu}$ is the extension by zero in $\mathbb{R} \setminus I$ of the measure ν given by (5.4.19).

Putting

$$\Phi(x) := \left(\frac{\rho}{\varphi}\right)^{p^*}(x), \qquad x \in I, \tag{5.4.31}$$

we have (cf. (5.4.5))

$$\Phi \in Q_{\rho^{p^*}}(I) \quad \text{and} \quad \{x_k\}_{k=K_-}^{K_+} \in CS\left(\Phi, \rho^{p^*}, I, \alpha^{p^*}\right). \tag{5.4.32}$$

By (5.4.30) and (5.4.31),

$$B^{p^*} = \left\| \int_{I_k} \Phi(t)\, d\overline{\nu} \right\|_{\ell^{r/p^*}(\mathcal{K}+)},$$

and thus, by Remark 4.2.6 and (5.4.32),

$$B^{p^*} \approx \left\| \frac{\Phi(x_k)}{\rho^p(x_k)} f(x_k) \right\|_{\ell^{r/p^*}(\mathcal{K}_-^+)}, \tag{5.4.33}$$

where

$$f(x) := \left\| \min\{\rho^{p^*}(\cdot), \rho^{p^*}(x)\} \right\|_{1, I, \nu}, \quad x \in I, \tag{5.4.34}$$

(i.e., f is the ρ^{p^*}-fundamental function of the space $L^1(I, \nu)$). If

$$\{y_k\}_{k=K_-^1}^{K_+^1} \in CS\left(f, \rho^{p^*}, I, \beta\right) \quad \text{with} \quad \beta > 1,$$

then, by (5.4.33) and Lemma 4.2.9 (with $f_1 = \Phi, f_2 = f$), we obtain

$$B^{p^*} \approx \left\| \frac{\Phi(y_k)}{\rho^{p^*}(y_k)} f(y_k) \right\|_{\ell^{r/p^*}(\mathcal{K}_-^1+)} \tag{5.4.35}$$

with

$$\mathcal{K}_-^{1+} := \{k \in \mathbb{Z}; K_-^1 \le k \le K_+^1\}.$$

Equation (5.4.22) still holds, and thus, on using Theorem 2.4.4, we obtain (5.2.66), which implies that, for all $x \in I$,

$$\begin{aligned}(5.4.36)\qquad \Phi(x) = \left(\frac{\rho}{\varphi}\right)^{p^*}(x) &= \left[\left(\frac{\rho^q}{H}\right)^{r/q}(x)\right]^{p^*/r} \\ &\approx \lim_{t\to a+}\left(\frac{\rho}{\varphi}\right)^{p^*}(t) + \left(\lim_{t\to b-}\frac{1}{\varphi^{p^*}(t)}\right)\rho^{p^*}(x) \\ &\quad + \|\min\{\rho^{p^*}(\cdot), \rho^{p^*}(x)\}\|_{r/p^*,W,I,\mu},\end{aligned}$$

where

$$(5.4.37)\qquad W(t) = \frac{\left(\int_a^t \rho^q(s)\,w^q(s)\,ds\right)^{p^*/r}\left(\int_t^b w^q(s)\,ds\right)^{p^*/r}}{\varphi(t)^{p^*+2p^*q/r}}$$

and

$$(5.4.38)\qquad d\mu(t) = d\rho^q(t).$$

Thus, on using (5.4.36) in (5.4.35), we arrive at

$$\begin{aligned}(5.4.39)\qquad B^{p^*} &\approx \left(\lim_{t\to a+}\left(\frac{\rho}{\varphi}\right)^{p^*}(t)\right)\left\|\frac{f(y_k)}{\rho^{p^*}(y_k)}\right\|_{\ell^{r/p^*}(\mathcal{K}_-^{1+})} \\ &\quad + \left(\lim_{t\to b-}\frac{1}{\varphi^{p^*}(t)}\right)\|f(y_k)\|_{\ell^{r/p^*}(\mathcal{K}_-^{1+})} \\ &\quad + \left\|\left\|\min\{\rho^{p^*}(\cdot), \rho^{p^*}(y_k)\}\right\|_{r/p^*,W,I,\mu}\frac{f(y_k)}{\rho^{p^*}(y_k)}\right\|_{\ell^{r/p^*}(\mathcal{K}_-^{1+})} \\ &=: A_1 + A_2 + A_3\end{aligned}$$

(recall that Convention 1.1.1 (i) is used when the expressions $0 \cdot (+\infty)$ appear in (5.4.39)). Consequently, condition (5.4.15) is equivalent to

$$(5.4.40)\qquad A_1^{1/p^*} + A_2^{1/p^*} + A_3^{1/p^*} < +\infty.$$

Since the sequence $\{f(y_k)/\rho^{p^*}(y_k)\}_{k\in\mathcal{K}_-^{1+}}$ is geometrically decreasing and the sequence $\{f(y_k)\}_{k\in\mathcal{K}_-^{1+}}$ is geometrically increasing, it follows from Remark 1.3.2 and Lemma 1.3.3 that

$$A_1 \approx \lim_{t\to a+}\left(\frac{\rho}{\varphi}\right)^{p^*}(t)\lim_{t\to a+}\frac{f(t)}{\rho^{p^*}(t)} \qquad\text{and}\qquad A_2 \approx \lim_{t\to b-}\frac{1}{\varphi^{p^*}(t)}\lim_{t\to b-} f(t).$$

Moreover, using (5.4.4) and Lemma 2.3.2, we arrive at

$$\lim_{t\to a+}\left(\frac{\rho}{\varphi}\right)^{p^*}(t) = \frac{1}{\|w\|_{q,I}^{p^*}} \qquad\text{and}\qquad \lim_{t\to b-}\frac{1}{\varphi^{p^*}(t)} = \frac{1}{\|\rho w\|_{q,I}^{p^*}}.$$

Similarly, making use of (5.4.34), (5.4.19) and Lemma 2.3.2, we obtain

$$\lim_{t\to a+}\frac{f(t)}{\rho^{p^*}(t)}=\nu(I)=\int_I \rho^{-p^*}(t)\,v^{p^*}(t)\,dt$$

and

$$\lim_{t\to b-} f(t)=\|\rho^{p^*}\|_{1,I,\nu}=\|v\|_{p^*,I}^{p^*}.$$

Consequently,

$$A_1^{1/p^*}\approx\frac{\|v/\rho\|_{p^*,I}}{\|w\|_{q,I}}\qquad\text{and}\qquad A_2^{1/p^*}\approx\frac{\|v\|_{p^*,I}}{\|\rho w\|_{q,I}}. \tag{5.4.41}$$

If

$$\{z_k\}_{k=K_-^2}^{K_+^2}\in CS\left(\left\|\min\{\rho^{p^*}(\cdot),\rho^{p^*}(x)\}\right\|_{r/p^*,W,I,\mu},\rho^{p^*},I,\beta_1\right)\quad\text{with}\quad\beta_1>1,$$

then an application of Lemma 4.2.9 followed by Theorem 4.2.5 gives

$$\begin{aligned}A_3&\approx\left\|\left\|\min\{\rho^{p^*}(\cdot),\rho^{p^*}(z_k)\}\right\|_{r/p^*,W,I,\mu}\frac{f(z_k)}{\rho^{p^*}(z_k)}\right\|_{\ell^{r/p^*}(\mathcal{K}_-^{2+})}\\&\approx\|f\|_{r/p^*,W,I,\mu},\end{aligned} \tag{5.4.42}$$

where

$$\mathcal{K}_-^{2+}:=\{k\in\mathbb{Z};K_-^2\le k\le K_+^2\}.$$

On collecting together (5.4.41) and (5.4.42), (5.4.37), (5.4.38),(5.4.34) and (5.4.19), one can easily verify that condition (5.4.40) coincides with (5.4.3).

Suppose now that $p=1$ and hence $p^*=\infty$. In this case (5.4.15) becomes

$$B:=\left\|\left\|\frac{\rho}{\varphi}\right\|_{\infty,v/\rho,I_k}\right\|_{\ell^r(\mathcal{K}_+)}<+\infty. \tag{5.4.43}$$

Putting

$$\Phi(x):=\left(\frac{\rho}{\varphi}\right)(x)\qquad x\in I, \tag{5.4.44}$$

we have (cf. (5.4.5))

$$\Phi\in Q_\rho(I)\quad\text{and}\quad\{x_k\}_{k=K_-}^{K_+}\in CS\,(\Phi,\rho,I,\alpha). \tag{5.4.45}$$

By (5.4.43) and (5.4.44),

$$B=\left\|\,\|\Phi\|_{\infty,v/\rho,I_k}\right\|_{\ell^r(\mathcal{K}_+)},$$

and thus, by Remarks 4.2.4 (i), (iii) and (5.4.45),

$$B \approx \left\| \frac{\Phi(x_k)}{\rho(x_k)} f(x_k) \right\|_{\ell^r(\mathcal{K}_-^+)}, \tag{5.4.46}$$

where

$$f(x) := \|\min\{\rho(\cdot), \rho(x)\}\|_{\infty, v/\rho, I}, \quad x \in I, \tag{5.4.47}$$

(i.e., f is the ρ-fundamental function of the space $L^\infty(I, v/\rho)$). If

$$\{y_k\}_{k=K_-^1}^{K_+^1} \in CS(f, \rho, I, \beta) \quad \text{with} \quad \beta > 1,$$

then, by (5.4.46) and Lemma 4.2.9 (with $f_1 = \Phi, f_2 = f$), we obtain

$$B \approx \left\| \frac{\Phi(y_k)}{\rho(y_k)} f(y_k) \right\|_{\ell^r(\mathcal{K}_-^{1+})} \tag{5.4.48}$$

with

$$\mathcal{K}_-^{1+} := \{k \in \mathbb{Z}; K_-^1 \leq k \leq K_+^1\}.$$

Equation (5.4.22) still holds, and thus, by Theorem 2.4.4, we obtain (5.2.67). In our case (5.2.67) becomes

$$\begin{aligned} \Phi(x) = \left(\frac{\rho}{\varphi}\right)(x) &= \left[\left(\frac{\rho^q}{H} \right)^{r/q} (x) \right]^{1/r} \\ &\approx \lim_{t \to a+} \left(\frac{\rho}{\varphi} \right)(t) + \left(\lim_{t \to b-} \left(\frac{1}{\varphi(t)} \right) \right) \rho(x) \\ &+ \quad \|\min\{\rho(\cdot), \rho(x)\}\|_{r, W, I, \mu} \quad \text{for all} \quad x \in I, \end{aligned} \tag{5.4.49}$$

where

$$W(t) = \frac{\left(\int_a^t \rho^q(s) w(s)^q\, ds \right)^{1/r} \left(\int_t^b w^q(s)\, ds \right)^{1/r}}{\varphi(t)^{1+2q/r}} \quad \text{and} \quad d\mu(t) = d\rho^q(t).$$

Therefore, on using (5.4.49) in (5.4.48), we arrive at

$$\begin{aligned} B &\approx \left(\lim_{t \to a+} \left(\frac{\rho}{\varphi} \right)(t) \right) \left\| \frac{f(y_k)}{\rho(y_k)} \right\|_{\ell^r(\mathcal{K}_-^{1+})} \\ &\quad + \left(\lim_{t \to b-} \frac{1}{\varphi(t)} \right) \|f(y_k)\|_{\ell^r(\mathcal{K}_-^{1+})} \\ &\quad + \left\| \|\min\{\rho(\cdot), \rho(y_k)\}\|_{r, W, I, \mu} \frac{f(y_k)}{\rho(y_k)} \right\|_{\ell^r(\mathcal{K}_-^{1+})} \\ &=: A_1 + A_2 + A_3 \end{aligned} \tag{5.4.50}$$

(recall that Convention 1.1.1 (i) is used when the expressions $0 \cdot (+\infty)$ appear in (5.4.50)). Consequently, condition (5.4.43) is equivalent to

$$A_1 + A_2 + A_3 < +\infty. \tag{5.4.51}$$

Since the sequence $\{f(y_k)/\rho(y_k)\}_{k\in\mathcal{K}_-^{1+}}$ is geometrically decreasing and the sequence $\{f(y_k)\}_{k\in\mathcal{K}_-^{1+}}$ is geometrically increasing, it follows from Remark 1.3.2 and Lemma 1.3.3 that

$$A_1 \approx \lim_{t\to a+} \Big(\frac{\rho}{\varphi}\Big)(t) \lim_{t\to a+} \frac{f(t)}{\rho(t)} \qquad \text{and} \qquad A_2 \approx \lim_{t\to b-} \frac{1}{\varphi(t)} \lim_{t\to b-} f(t)\,.$$

Moreover, using (5.4.4) and Lemma 2.3.2, we get

$$\lim_{t\to a+} \Big(\frac{\rho}{\varphi}\Big)(t) = \frac{1}{\|w\|_{q,I}} \qquad \text{and} \qquad \lim_{t\to b-} \frac{1}{\varphi(t)} = \frac{1}{\|\rho w\|_{q,I}}\,.$$

Similarly, making use of (5.4.47) and Lemma 2.3.2, we obtain

$$\lim_{t\to a+} \frac{f(t)}{\rho(t)} = \Big\|\frac{v}{\rho}\Big\|_{\infty,I} \qquad \text{and} \qquad \lim_{t\to b-} f(t) = \Big\|\rho\,\frac{v}{\rho}\Big\|_{\infty,I} = \|v\|_{\infty,I}\,.$$

Consequently,

$$A_1 \approx \frac{\big\|\frac{v}{\rho}\big\|_{\infty,I}}{\|w\|_{q,I}} \qquad \text{and} \qquad A_2 \approx \frac{\|v\|_{\infty,I}}{\|\rho w\|_{q,I}}\,. \tag{5.4.52}$$

If

$$\{z_k\}_{k=K_-^2}^{K_+^2} \in CS\Big(\|\min\{\rho(\cdot),\rho(x)\}\|_{r,W,I,\mu}\,, \rho, I, \beta_1\Big) \quad \text{with} \quad \beta_1 > 1,$$

we apply Lemma 4.2.9, then Theorem 4.2.3 and Remarks 4.2.4 to derive

$$A_3 \approx \Big\| \|\min\{\rho(\cdot),\rho(z_k)\}\|_{r,W,I,\mu} \frac{f(z_k)}{\rho(z_k)} \Big\|_{\ell^r(\mathcal{K}_-^{2+})} \approx \|f\|_{r,W,I,\mu}\,, \tag{5.4.53}$$

where

$$\mathcal{K}_-^{2+} := \{k \in \mathbb{Z}; K_-^2 \le k \le K_+^2\}.$$

Finally, using (5.4.52), (5.4.53) and (5.4.47), one can see that condition (5.4.51) coincides with (5.4.3). □

The case $q = +\infty$ was not considered in Theorem 5.4.1. Note that if $q = +\infty$, then $r = p$. On taking into account Remark 5.2.7, the following theorem can be proved analogously to Theorem 5.4.1, Case B, with the help of the following observations.

1) If $q = +\infty$ and $p \in (0,1)$, then, instead of (5.4.36), one makes use of (5.2.76), which implies that, for all $x \in I$,

$$\Phi(x) := \Big(\frac{\rho}{\varphi}\Big)^{p^*}(x) = \Big[\Big(\frac{\rho}{\varphi}\Big)^{p}(x)\Big]^{p^*/p}$$
$$\approx \lim_{t\to a+} \Big(\frac{\rho}{\varphi}\Big)^{p^*}(t) + \Big(\lim_{t\to b-} \frac{1}{\varphi^{p^*}(t)}\Big)\rho^{p^*}(x)$$
$$+ \|\min\{\rho^{p^*}(\cdot), \rho^{p^*}(x)\}\|_{p/p^*,W,I,\mu},$$

where

$$W(t) = \Big(\frac{\mathcal{V}(t)}{\varphi(t)^{p+2}}\Big)^{p^*/p}, \quad d\mu(t) = d\rho(t), \quad t \in I,$$

and $\mathcal{V}$ defined in (5.2.77).

2) Similarly, if $q = +\infty$ and $p = 1$, then, instead of (5.4.49), one makes use of (5.2.76), which implies that, for all $x \in I$,

$$\Phi(x) := \Big(\frac{\rho}{\varphi}\Big)(x) = \Big[\Big(\frac{\rho}{\varphi}\Big)^{p}(x)\Big]^{1/p}$$
$$\approx \lim_{t\to a+} \Big(\frac{\rho}{\varphi}\Big)(t) + \Big(\lim_{t\to b-} \Big(\frac{1}{\varphi(t)}\Big)\Big)\rho(x)$$
$$+ \quad \|\min\{\rho(\cdot), \rho(x)\}\|_{p,W,I,\mu},$$

where

$$W(t) = \Big(\frac{\mathcal{V}(t)}{\varphi(t)^{p+2}}\Big)^{1/p}, \qquad d\mu(t) = d\rho(t), \quad t \in I,$$

and $\mathcal{V}$ is again given in (5.2.77).

Theorem 5.4.4 *Let $I = (a,b) \subseteq \mathbb{R}$, $0 < p \le 1$, $w, v, u \in \mathcal{W}(I)$. Moreover, let the fundamental function $\varphi(x) = \|\min\{\rho(\cdot), \rho(x)\}\|_{\infty,w,I}$, $x \in I$, satisfy (5.2.75) for some non-negative Borel measure $\widetilde{\mu}$. Suppose that $\rho(x) := \int_a^x u(t)\,dt$, $x \in I$, is such that $\rho \in Ads(I)$. Then*

$$\|g\|_{p,v,I} \lesssim \Big\| \int_a^x u(t) \Big(\int_t^b g(s)\,ds \Big) dt \Big\|_{\infty,w,I} \quad \textit{for all } g \in \mathcal{M}^+(I), \tag{5.4.54}$$

if and only if

$$\Big(\int_I \Big\| \min\Big\{1, \frac{\rho(x)}{\rho(\cdot)}\Big\}\Big\|_{p^*,v,I}^{p} \frac{\mathcal{V}(t;\alpha,\beta,\widetilde{\mu},\rho)}{\|\min\{\rho(\cdot),\rho(x)\}\|_{\infty,w,I}^{p+2}}\, d\rho(x)\Big)^{1/p} + \frac{\|v/\rho\|_{p^*,I}}{\|w\|_{\infty,I}} + \frac{\|v\|_{p^*,I}}{\|\rho w\|_{\infty,I}} < +\infty, \tag{5.4.55}$$

where the function $\mathcal{V}(\,\cdot\,;\alpha,\beta,\widetilde{\mu},\rho)$ is given in (5.2.77).

Bibliography

[BeSha] C. Bennett and R. Sharpley. Interpolation of operators, Pure and Applied Mathematics, vol. 129, Academic Press, New York, 1988.

[BK] Yu. A. Brudnyĭ and N. Ya. Krugljak. Interpolation Functors and Interpolation Spaces, Volume 1, North-Holland, Amsterdam, 1991.

[BL] J. Bergh and J. Löfström. Interpolation Spaces. An Introduction, Springer-Verlag, Berlin, 1976.

[BrSh] Yu. A. Brudnyĭ and A. Shteinberg. Calderón couples of Lipschitz spaces. *J. Funct. Anal.* **131** (1995), 459–498.

[C] M. Cwikel. K-divisibility of the K-functional and Calderón couples. *Ark. Mat.* **22** (1984), no. 1, 39–62.

[EGO1] W. D. Evans, A. Gogatishvili and B. Opic. The reverse Hardy inequality with measures. *Math. Inequal. Appl.* **11** (2008), no. 1, 43–74.

[EGO2] W. D. Evans, A. Gogatishvili and B. Opic. The ρ-quasiconcave Functions and Weighted Inequalities. In C. Bandle, A. Gilányi, L. Losonczi, Z. Páles and M. Plum (Editors), Inequalities and Applications International Series of Numerical Mathematics, vol. 157, Birkhäuser Verlag AG, Basel-Boston-Berlin, 2009, pp. 121–132.

[Fe] H. Federer. Geometric Measure Theory, Springer-Verlag, Berlin, 1969.

[Fo] G. B. Folland. Real Analysis, John Wiley, Inc., New York, 2nd ed., 1999.

[GJOP] A. Gogatishvili, M. Johansson, C. A. Okpoti and L.-E. Persson. Characterisation of embeddings in Lorentz spaces. *Bull. Austral. Math. Soc.*, **76** (2007), no. 1, 69–92.

[GMP1] A. Gogatishvili, R. Mustafayev and L.-E. Persson. Some new iterated Hardy-type inequalities: the case $\theta = 1$. *J. Inequal. Appl.*, 2013, 2013:515, 29 pp.

[GMP2] A. Gogatishvili, R. Mustafayev and L.-E. Persson. Some new iterated Hardy-type inequalities. *J. Funct. Spaces Appl.*, 2012, Art. ID 734194, 30 pp.

[GOP] A. Gogatishvili, B. Opic and L. Pick. Weighted inequalities for Hardy-type operators involving suprema. *Collect. Math.*, **57** (2006), no. 3, 227–255.

[GP1] A. Gogatishvili and L. Pick. Discretization and anti-discretization of rearrangement invariant norms. *Publ. Mat.*, **47** (2003), 311–358.

[GP2] A. Gogatishvili and L. Pick. Embeddings and duality theorems for weak classical Lorentz spaces. *Canad. Math. Bull.* , **49** (2006), no. 1, 82–95.

[G1] M. L. Gol'dman. On integral inequalities on a cone of functions with monotonicity properties. *Soviet Math. Dokl.*, **44** (1992), 581–587.

[G2] M. L. Gol'dman. On integral inequalities on the set of functions with some properties of monotonicity, in: *Function spaces, differential operators and nonlinear analysis* (Friedrichroda, 1992), Teubner-Texte Math. **133**, Teubner, Stuttgart, (1993), 274–279.

[GHS] M. L. Gol'dman, H. P. Heinig and V. D. Stepanov. On the principle of duality in Lorentz spaces. *Can. J. Math.*, **48** (1996), 959–979.

[HLP] G. Hardy, J. E. Littlewood and G. Pólya. Inequalities, Cambridge University Press, Cambridge, 1934 (2nd ed. 1952).

[J] S. Janson. Minimal and maximal method of interpolation. *J. Funct. Anal.*, **44** (1981), 50–73.

[K] M. Křepela. Integral conditions for Hardy-type operators involving suprema. *Collect. Math.*, **68** (2017), 21–50.

[KP] A. Kufner and L.-E. Persson. Weighted inequalities of Hardy type, World Scientific Publishing Co., Inc., River Edge, NJ, 2003.

[L1] L. Leindler. Inequalities of Hardy-Littlewood. *Anal. Math.*, **2** (1976), 117–123.

[L2] L. Leindler. On the converses of inequalities of Hardy and Littlewood. *Acta Sci. Math. (Szeged)*, **58** (1993), 191–196.

[N] P. Nilsson. Interpolation of Banach lattices. *Studia Math.*, **82** (1985), 135–154.

[OPS] C. Okpoti, L.-E. Persson and G. Sinnamon. An equivalence theorem for some integral conditions with general measures related to Hardy's inequality. *Math. Anal. Appl.*, **326** (2007), 398–413.

[OK] B. Opic and A. Kufner. Hardy-Type Inequalities, Pitman Res. Notes in Math. Ser. 219, Longman Sci. & Tech., Harlow, 1990.

[Osk1] K. I. Oskolkov. Estimation of the rate of approximation of a continuous function and its conjugate by Fourier sums on a set of full measure. (Russian) *Izv. Akad. Nauk SSSR Ser. Mat.*, **38** (1974), 1393 -1407.

[Osk2] K. I. Oskolkov. Approximation of integrable functions on sets of full measure. (Russian) *Mat. Sb. (N.S.)*, **103 (145)** (1977), no. 4, 563–589.

[Ov1] V. I. Ovchinnikov. The method of orbits in interpolation theory. Math. Rep. **1** (1984), no. 2, i-x + 349–515.

[Ov2] V. I. Ovchinnikov. Interpolation functions and the Lions-Peetre interpolation construction. (Russian) *Uspekhi Mat. Nauk*, **69** (2014), no. 4(418), 103–168; translation in *Russian Math. Surveys* **69** (2014), no. 4, 681–741.

[P] L.-E. Persson. Relations between summability of functions and their Fourier series. *Acta Math. Acad. Sci. Hungar.*, **27** (1976), no. 3–4, 267–280.

[Pr1] D. V. Prokhorov. Boundedness and compactness of a supremum-involving integral operator. *Proc. Steklov Inst. Math.*, **283** (2013), no. 1, 136–148.

[Pr2] D. V. Prokhorov. Lorentz norm inequalities for the Hardy operator involving suprema. *Proc. Amer. Math. Soc.* **140** (2012), no. 5, 1585–1592.

[PS1] D. V. Prokhorov and V. D. Stepanov. On supremum operators. (Russian) *Dokl. Akad. Nauk* **439** (2011), no. 1, 28–29; translation in *Dokl. Math.* **84** (2011), no. 1, 457–458.

[PS2] D. V. Prokhorov and V. D. Stepanov. Weighted inequalities for quasilinear integral operators on the semiaxis and applications to Lorentz spaces. (Russian) *Mat. Sb.*, **207** (2016), no. 8, 135–162; translation in *Sb. Math.* **207** (2016), no. 7–8, 1159–1186.

[RR] M. M. Rao and Z. D. Ren. Theory of Orlicz Spaces, Marcel Dekker Inc., New York, 1991.

[R] W. Rudin. Principles of Mathematical Analysis, McGraw-Hill Book Company, New York, 2nd ed., 1964.

Index

Printed in the United States
By Bookmasters